The Hidden Energy Crisis

Praise for this book...

'This book provides a well needed insight into the challenges facing the developing world and the strong link between poverty and energy provision. Sanchez argues that progress in providing the minimum energy requirements needed to raise the Human Development Index has been very slow and there is little to suggest much progress will be made in reducing the 1.62 billion people currently without electricity. The dimension of climate change and its effects are addressed which is vital given that this aspect will affect the world's poorest more than any other sector. Sanchez gives a number of suggestions concerning how policies of the major funding agencies should be modified in light of arguments presented. Essential reading for anyone working in the area of energy in the context of international development.'

Dr Keith R. Pullen, Reader in Energy Systems,
City University, London

'Teo Sanchez touches a raw nerve in the world of energy access – the majority of the poor who remain peripheral to modern energy interventions. Unless we provide a visible face of the poor, energy access will continue to remain low on the agenda of national government policies or international aid. Sanchez starts the debate with facts and figures, which I hope will convert into positive actions.'

Dr Kavita Rai, Programme Manager,
Global Village Energy Partnership

The Hidden Energy Crisis
How policies are failing the world's poor

Teodoro Sanchez

PRACTICAL ACTION
Publishing

Practical Action Publishing Ltd
Schumacher Centre for Technology and Development
Bourton on Dunsmore, Rugby,
Warwickshire, CV23 9QZ, UK
www.practicalactionpublishing.org

© Practical Action Publishing, 2010

ISBN 978 1 85339 676 2

Since 1974, Practical Action Publishing (formerly Intermediate Technology
Publications and ITDG Publishing) has published and disseminated books
and information in support of international development work throughout
the world. Practical Action Publishing Ltd (Company Reg. No. 1159018) is
the wholly owned publishing company of Practical Action. Practical Action
Publishing trades only in support of its parent charity objectives and any
profits are covenanted back to Practical Action (Charity Reg. No. 247257,
Group VAT Registration No. 880 9924 76).

Photo: Child sitting next to a kerosene lamp, Nepal Credit: Lindel Caine
Indexed by Andrea Palmer
Typeset by S.J.I. Services
Printed by Hobbs The Printers

Contents

Boxes, figures and tables

Boxes

Figures

Tables

Acronyms

CDM	Clean Development Mechanism
CEIF	Clean Energy for Development Investment Framework
CSD	Commission on Sustainable Development
DFID UK	Department for International Development
ECOSOC	Economic and Social Council of the United Nations
GATS	General Agreement on Trade in Services
GDP	Gross Domestic Product
GEF	Global Environment Facility
GNESD	Global Network for Sustainable Development
GVEP	Global Village Energy Partnership
GW	gigawatt
HDI	Human Development Index
IEA	International Energy Agency
kW	kilowatt
kWh	kilowatt hour
LPG	liquefied petroleum gas
MDGs	Millennium Development Goals
ODA	official development assistance
OECD	Organization for Economic Co-operation and Development
PV	photovoltaic systems
REA	Rural Electrification Administration
REEEP	Renewable Energy and Energy Efficiency Partnership
SHS	solar home systems
TOE	ton oil equivalent
UNDP	United Nations Development Programme
UNESCO	United Nations Educational, Scientific and Cultural Organization
UNFCCC	United Nations Framework Convention on Climate Change
UNIDO	United Nations Industrial Development Organization
W	watt
W_p	watt peak
WTO	World Trade Organization

About the author

Teodoro Sanchez holds the position of Energy Technology and Policy Adviser in Practical Action in the UK and in this capacity takes part in a number of international projects regarding policy and technology for energy access for the poor in developing countries. Born in the Peruvian Andes in 1953, he graduated as a Mechanical Engineer at the National University of Engineering (UNI) in Peru in 1978, obtained an MSc degree in Renewable Energies at the University of Reading UK in 1988 with the support of a British Council Scholarship, and completed his PhD studies in 2006 at Nottingham Trent University with the theme 'The critical success factors for isolated stand-alone energy schemes in Peru'.

His professional career has been devoted to small-scale renewable energies and to rural electrification in developing countries. From 1978 to 1988 he worked as a research engineer at the Energy Division of ITINTEC (National Institute for Technological Research and National Standards, Peru) and from 1988 to the present day with Practical Action (formerly ITDG), first in Peru and since 2004 in the UK. At Practical Action his responsibilities have ranged from field engineer to Programme Manager and then to Energy Technology and Policy Adviser, and he has been involved in different national and international assignments related to energy technology and policy in developing countries within Latin America, Africa and Asia. He has also carried out consultancy work for several international organizations such as The World Bank, UNDP, EU and other multilateral and bilateral organizations. He is an author and co-author of several articles and books related to small-scale renewable energies and rural electrification. He has also been a lecturer at the Post-Graduate School of the UNI in Peru between 1989 and 2004, teaching courses on small-scale renewable energy technologies and rural electrification.

Foreword

The economic recession, which dominated headlines in the first half of 2009, has temporarily reduced the demand for energy across the globe. As a result, oil prices are lower than a year ago and the rate of growth of greenhouse gas emissions has been tempered. Nevertheless, energy supplies and energy consumption continue to be a major feature in public policy debate, internationally and nationally. The same headlines in 2008, before the financial crisis, were dominated by the 'energy crisis' and the high price of oil, and concerns about climate change are now never out of the news. There is little doubt that energy security and climate change will be key policy issues for some time to come, at risk of crowding out or overshadowing debate about the other hidden energy crisis which is the subject of this book.

Energy security has become a more significant factor in domestic and international policy because of fluctuating oil prices, climate change, and concerns about security and terrorism more generally. Developing countries are particularly affected by variability in oil prices and the new scramble for energy resources by energy importing countries. The energy security of both parties is now often on the agenda when development cooperation is discussed between donors and recipients. But, unlike debates about food security, the energy security of people and families living in poverty is not widely discussed, even in international development circles.

Energy is a principal source of greenhouse gas emissions, so inevitably energy policy is linked to climate change policy. For climate change mitigation reasons there has been a tendency to exclude consideration of some energy technology options for developing countries solely on the basis of their carbon emissions. Even the choice of technology for cooking by women in poor households has been highlighted as a climate change issue because of the effects of 'black carbon'. Yet, the increase in consumption of modern energy services necessary to enable poor people in developing countries to escape poverty would not add significantly to global emissions.

Poverty reduction does mean increased energy consumption. Though the energy needs of people living in poverty are small, and small amounts of energy can make a significant difference to their lives, millions of men, women and children in the developing world do continue to live in absolute poverty because they do not have access to modern energy services. Though millions of people, mostly in Asia, have in recent years managed to escape poverty and have secured access to modern energy services, achievement of each of the Millennium Development Goals (MDGs) continues to be constrained by energy poverty across the developing world. On current trends, by 2030, the number of people without access to modern energy services will be the same as today.

Yet, simple and effective technologies are available to deliver clean and efficient energy services to people living in resource-poor communities, whether they are in remote rural locations or peri-urban settings. Alongside conventional means of delivery, such as grid extension or fossil fuels, decentralized technologies which use local energy resources, for example micro-hydropower, solar power and improved cooking stoves, can be used effectively to supply the energy necessary for poverty reduction. The barriers to realizing universal access to modern energy services appear not to be technological, but the availability of funds to invest in pro-poor energy services and the political will to ensure that such investment takes place.

Official development assistance (ODA) for the energy sector accounted for about 4 per cent of total ODA worldwide in 2006 – US$4,958 million. However, ODA for the energy sector includes investment for the extraction of energy resources; fossil fuels. Investment in energy services may also be undertaken through cross-sectoral programmes and thus not be classified as energy sector ODA. Surprisingly, therefore, it is quite difficult to obtain an overall picture about aid commitments for energy access or investment in energy services for people living in poverty.

What we can say is that ODA for energy in developing countries has generally been targeted at delivering economic growth. National and regional policies for the energy sector are similarly oriented towards economic growth. This has resulted in a focus on investment in large-scale energy infrastructure (i.e. large-scale coal, large hydro, transmission grid, pipelines), including infrastructure for the export of energy to industrialized countries or urban centres. For instance, half of the €664 million invested in the energy sector of the African, Caribbean and Pacific Group of States (ACP countries) by the European Commission, during the decade to 2006, was for power generation. By contrast cooking, the main energy demand for most people living in poverty in developing countries, rarely features in poverty reduction strategies and donor country assistance programmes. As argued by Teo Sanchez in this book, there is a need to redress the balance, with much more attention and investment being directed towards the supply of local energy services for poverty reduction in local communities.

If the donor community is less than informative about ODA commitments to energy services for people living in poverty, donors probably have even less to say about the impact on poverty of their investments in energy. Evaluation of a decade of EC cooperation in the energy sector concluded that the effects of EC support on access to energy for the poor 'have been limited to date' (European Commission, 2008).

More tellingly, the impact of EC energy-related activities on living conditions and growth had simply not been measured. At the World Bank, the largest multilateral donor in the energy sector, an evaluation of their rural electrification (RE) projects found that 'poverty has not become a central concern of RE projects, and there is rarely any explicit consideration either of how the poor will be included or of any poor-specific activities' (World Bank, 2008). The same study found the evidence base to be weak for many of the benefits claimed for rural electrification, particularly health and education benefits. Overall, there seems to be a remarkable lack of knowledge about the poverty impact of aid to the energy sector.

One reason why there has been little systematic attempt to assess the impact on poverty reduction of energy-related ODA may be that there are no internationally agreed MDG style targets for energy. Yet, they have been suggested. The UN Millennium Project in 2004 noted that to achieve the MDGs the proportion of the world's population without access to basic electricity would have to be reduced and that the population reliant on traditional solid fuels for cooking should be reduced to no more than 1 billion people by 2015. The International Energy Agency has recommended that the number of people without electricity should be reduced to 1 billion, and the number of people reliant on traditional biomass reduced to 1.85 billion by 2015. At the 15th session of the UN Commission for Sustainable Development (CSD) in 2007 there were similar calls from civil society and other non-governmental organizations. However, there has been a complete failure to agree any international targets, or international programmes or action, towards ensuring access to energy services.

What would it take to enable everyone to have sufficient access to energy services? The Millennium Project estimated that a per capita investment in energy of $10–$20 would be needed to meet the MDGs. For the 1.6 billion without access to electricity this amounts to a total worldwide of $16–$32 billion. The World Bank estimated that to achieve 100 per cent access to electricity, in all regions of the world by 2030, $34 billion a year would be necessary. Compared to the small proportion of the $5 billion total energy aid, there remains a substantial difference between the amounts allocated for energy and the amount required even to ensure achievement of the MDGs. In *The Hidden Energy Crisis*, Teo Sanchez provides his own estimates for what is needed to provide adequate energy to everyone who currently does not have access to modern energy services.

The links between access to energy and poverty reduction have been recognized by the international community since the World Summit for Sustainable Development in 2002. Several multi-stakeholder partnerships were launched at WSSD in Johannesburg to strengthen and promote the contribution of energy to sustainable development and poverty reduction. And since 2002 there has been an increase in ODA for the energy sector overall. The commitment to energy access, however, is largely rhetorical. The European Commission appears to consider it as a climate change matter, rather than one of poverty reduction and the realization of basic rights. The World Bank is spending more on renewable energy, but not necessarily focused on energy services for people living in poverty. And the CSD in 2007 failed to agree a statement on energy for sustainable development, and in the absence of any other recognized international forum to debate energy policies and human development, this leaves the commitment of the international community open to question.

Everyone needs energy to cook food, to heat the home, to earn a living, to benefit from good health and education services. Lack of access to energy services denies people a basic standard of living which should be available for all. With funding and the political will to provide the funding being critical to eliminating energy poverty, there is a continuing need to make the case for change. This book provides a useful and timely contribution.

Andrew Scott, Practical Action

CHAPTER 1
Inequalities in energy

Energy is fundamental to human life

There is a long history of energy use by humankind, beginning when fire was discovered for cooking, heating and scaring away wild animals, long before people could read and write. Fire was civilization's first great energy discovery and wood was the main fuel for a long time. In developed countries people have moved from using wood as fuel for basic cooking to modern energy sources such as gas and electricity. However, a primitive hearth is still the reality today for most of the rural, and a large proportion of the urban, people living in developing countries.

Energy is critical for human development. It allows people to access a range of basic services, including drinking water, health, education, transport, communication and other essential services. However, currently a large proportion of the world's population is confronted with absolute energy poverty. They are without access to electricity and are obliged to cook and heat with solid biomass, such as wood, dung or charcoal, most of the time using inefficient devices and in unhealthy conditions.

For developed countries, the switch to modern energy cooking fuels was not straightforward and in most cases it required government investments in infrastructure, policies, strategies and subsidies (see chapter 2). Some developing countries also have policies to promote the use of liquid and gas fuels (modern fuels), however the reality is that few of these policies work for the poor. During the last three decades there have been important global efforts to reduce energy poverty, by governments as well as by international development agencies and financial institutions.

Definitions

Over the last two decades there have been a number of terms used to express the insufficiency or poor quality of energy services and sources for poor households. The most common have been 'energy poverty', 'energy access', and 'energy access for the poor'. However none of these terms have had a unique definition. For the purpose of this book the following definitions are adopted:

Energy poverty. People are considered to be in 'energy poverty' if they do not have access to at least the equivalent of 35 kg of liquefied petroleum gas (LPG) for cooking per capita per year from liquid fuels, gas or supplies of improved solid fuel, as well as efficient and clean cooking stoves.[1] In addition, access to 120 kWh of electricity per capita per year is a minimum requirement for lighting, access to basic services (such as drinking water, communications, health care and education etc.), as well as providing some added contribution towards local production.

Energy access. This is defined as access to a sufficient quantity of energy to fulfil basic services and access for income and development opportunities. It is clear that the quantity of energy needed to deliver the needs of a large number of people living in 'energy poverty' is very small compared to the total world demand for energy. The investments needed to make this possible are small compared with the huge sums of money needed for global energy security.

Energy access for the poor. This refers to the provision of energy services and energy sources to the poorest groups of people, marginalized either because of poverty or physical isolation (for example, living in marginal urban areas or away from towns in remote small villages and communities) or both. In rural areas there may be a small proportion of people who are not very poor in economic terms, but are energy poor because they have no access to energy to fulfil their basic needs.

Energy poverty and lack of access

Recent reports from leading energy agencies (IEA, 2006) show that about 1.62 billion people worldwide have no access to electricity services. The majority of these live in rural areas in developing countries. Nearly 3 billion people are forced to cook with solid fuels, often using low quality biomass sources – such as dung, agricultural residues and charcoal – or with low-grade coal, using very low efficiency cooking devices, which contribute to drudgery, poverty, illnesses and death.

Energy access for the poor is a huge challenge, in fact even the most optimistic forecasts consider energy poverty in developing countries will take many decades to overcome, especially energy access for the poorest sectors of the population. The International Energy Agency (IEA), for example, shows that despite the $10.5 trillion estimated required investment in developing countries in the period 2005–30 (IEA, 2006) there will still be about 1.4 billion people without access to electricity, which would mean a reduction in nearly 25 years of only 12.5 per cent of the current number of people without electricity. There is also predicted to be 2.6 billion still relying on traditional biomass for cooking by 2030. This explains the magnitude of the challenge, while other forecasts confirm that in less developed countries fuel wood and agricultural residues will be relied on for cooking for the foreseeable future.

The UNDP Fifth Africa Governance Forum (AGF-V) held in Maputo in 2002 defined poverty as a combination of many deprivations, including economic, human, political, socio-cultural, protective and cross-cutting gender-related and environmental hardship (UNDP, 2002). In this respect energy poverty is related to the absence of sufficient choice in accessing adequate, affordable, high-quality, safe and environmentally benign energy services to support economic and human development (UNDP and WEC, 2004).

Over the last few decades there have been important discussions regarding energy access, and several new terms relating to energy access needs have entered the debates. These include: rural energy, productive energy, energy access for the poor, energy poverty, among others. Most are used with the intention of putting emphasis on particular issues around energy. For example, rural energy highlights the issue that all energy needs in rural areas should be counted (such as cooking fuels, mechanical power, space heating and electricity) and productive energy introduces the idea that energy access has to focus on productive uses. The concept of productive energy is not new but gained prominence in the 1990s and was developed to persuade donors and energy promoters that energy will be good for the poor if it is related to income generation. These concepts have been useful to make planners and policymakers aware of the different needs of the poor, but it appears that these concepts have contributed little to reducing energy poverty.

Progress towards achieving energy access for the poor has been slow during the last decades as Saghir (2005: 8) states:

Advances over the past 25 years have been remarkable, with more than one billion people in developing countries gaining access to electricity and modern fuels. But as impressive as this accomplishment is, large gaps in access remain. While (mostly urban) higher-income households now have access to modern energy, the world's poorest (mostly rural) households do not. Some regions lag further than others. While problems of access are now far greater in rural areas than in urban areas, the rapid growth expected in urban populations in the next decades could lead to growing gaps in access to electricity.

The Population Division of the UN Department of Economic Affairs forecasts that urban populations in developing countries will increase by 1.65 billion between 2005 and 2030, and in rural areas by around 50 million people. This shift in population from rural to urban areas during the coming decades means increased pressure on delivering basic services in urban areas, including both electricity and cooking fuels.

The Human Development Report (UNDP, 2005), shows that about 1 billion people live on less than US$1 per day, and 2.5 billion live on less than $2. The report does not provide information on who (if any) have access to electricity and/or modern cooking fuels. However, there is no doubt that those with the lowest income are also the energy poor and most of them are unlikely to gain access to modern energy unless dramatic changes happen with regards to policies, strategies, allocation of resources and priorities. There are many questions unanswered regarding energy access for the poor, such as:

- What are the energy needs of the poor and the scale of the energy demand to meet these needs?

- What are the financial requirements (the funding gap) and where will the money come from to enable the poor to gain access to energy?
- Apart from the financial gap, what are the other challenges to accessing energy both in rural and urban areas?

This book attempts to arrive at conclusions and answer some of these concerns. It also proposes an agenda for greater access of energy for the poor.

Energy and development

Literature includes a wide range of information showing that energy is a critical element for human development. Energy is needed to provide access to, or to improve the quality of, a range of other basic services such as health, education, drinking water, communications, transport and others. Energy is required to improve agricultural production and animal husbandry, for building and shelter, as well as to power equipment for the processing of harvested products. Energy is also critical for household activities, such as cooking, where traditionally women perform the main role. The provision of cleaner and healthier fuels is essential to help free-up women's time and to avoid health problems for women and children.

The World Bank, UNDP, European Union (EU) and other high level organizations show that energy access has clear links with Gross Domestic Product (GDP) and human development indicators, as will be seen in the following sections. Countries with the lowest electricity access and with high biomass energy consumption also have the lowest GDP and lowest Human Development Index (HDI).

Energy access for the poor and the Millennium Development Goals

Recently, organizations such as UNDP, the World Bank and the European Economic Community (EEC) have recognized the strong link between energy access and seven of the MDGs. Most now agree that achievement of the MDGs are heavily dependent on energy access for the poor. However, when setting the MDG targets and indicators, these organizations failed to consider the vital role energy plays in the achievement of MDGs. In total there are eight MDGs, with 18 targets and 48 indicators. Only one of the MDGs (MDG 7: Ensure environmental sustainability) indirectly considers energy at all. The issue of energy is hidden under the target: 'Integrate the principles of sustainable development into country policies and programmes and reverse the loss of environmental resources'. This target has three indicators: addressing the national averages on energy consumption, per capita CO_2 emissions and the proportion of people using fossil fuels. However even with these indicators, the environmental concerns expressed do not emphasize the importance of energy access for the poor but that of cutting greenhouse gas emissions.

In 2002, governments and leaders at the Johannesburg Sustainable Development Summit failed to agree on a goal or clear targets for energy access for the poor. Although UNDP claims that energy for sustainable development was given prominence at the summit, the truth is that energy access for the poor was given little attention and there was a lack of awareness on the part of UNDP and ECOSOC (The Economic and Social Council of the United Nations) of the importance of energy in relation to poverty.

With regards to what the MDGs have achieved so far, the World Bank shows that globally there has been some progress towards the goals. In its report *World Development Indicators* the World Bank (2005) states:

Since 1990 rates of extreme poverty have declined in many countries. The majority of the extremely poor population lives in countries that are on track to achieve the Millennium Development Goal target. This includes countries with large populations such as China, India, Pakistan, and

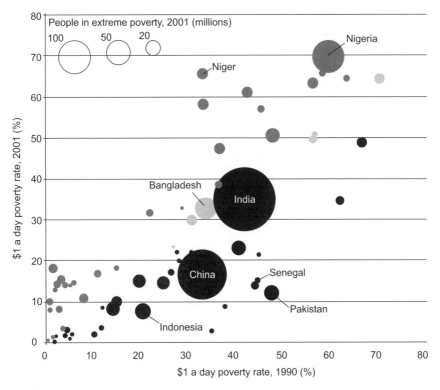

Figure 1.1 $1 a day poverty rate, 2001 (%) against 1990 levels
Source: World Bank, 2005

Indonesia, where many people have climbed out of poverty. In other countries, such as Bangladesh, poverty rates have been declining but not fast enough to be cut in half by 2015. And there are many more countries where poverty rates have increased since 1990. In Sub-Saharan Africa only a handful of countries such as Senegal are on track to reach the target. Reversing the trend will require higher rates of economic growth and benefits reaching the poor – a daunting task on top of the burdens of disease, famine, and armed conflict.

The same report (World Bank, 2005) also shows that most of the countries which are on track to achieve the MDGs (see Figure 1.1) are investing in important energy access programmes or they already have more widespread energy access, such as the case of China and India. However, countries with no investment in energy, such as the majority of the sub-Saharan African countries, are also off-track on the achievement of the MDGs.

Energy consumption as a development vector

An important issue regarding energy and human development is that energy access is fundamental to increasing the HDI in developing countries, especially within the poorest sectors of the population. However, in rich countries the increase of energy consumption no longer contributes to increasing the HDI any further. Figure 1.2 shows a clear correlation between HDI and the average annual energy consumption per capita for a number of countries, this

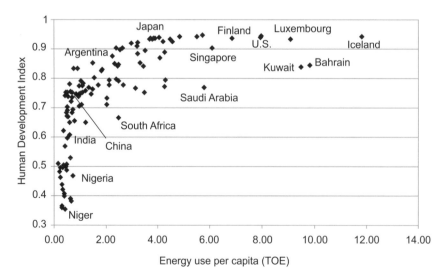

Figure 1.2 Correlation between energy consumption and HDI
Source: Sager et al., 2006

includes all energy consumption including electricity, industrial fuels, and cooking and heating fuels. This figure in fact shows that the correlation line reaches its plateau at about 3 TOE (ton oil equivalent, approximately 34,890 kWh) average per capita annual energy consumption, any larger average energy consumption does not contribute to increasing HDI. Furthermore Figure 1.2 also shows that an average energy consumption of 1 TOE (11,630 kWh) correlates with a HDI of around 0.75 or less.

Within calculations of total energy consumption, there have been investigations into the role specifically of electricity in development. The literature shows there is a good correlation between per capita electricity consumption and HDI. Pasternak (2000, see Figure 1.3) shows that HDI reaches a plateau when nations consume about 4,000 kWh per capita electricity annually, which implies that for poor countries presently consuming much less than that figure every small increase in energy consumption could improve HDI, especially in the lower range (below 1,000 kWh). While for rich countries consuming more than 4,000 kWh per capita an increase in electricity consumption makes little or no difference to their HDI. In these rich countries anything above this level is used in excess; it is either wasted or used inefficiently or both.

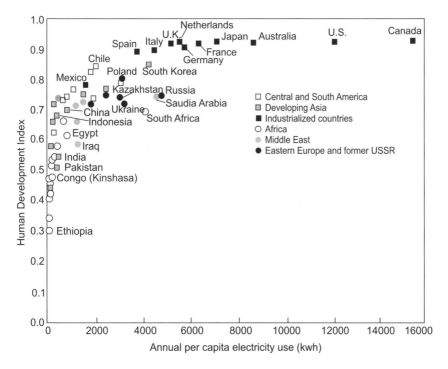

Figure 1.3 Correlation between electricity consumption and HDI
Source: adapted from Pasternak, 2000

Energy needs

The basic energy needs of humankind are heating, cooking, lighting and mechanical power (such as milling). The intensity of energy consumption and energy options differ depending on lifestyle, wealth, location and type of activities of the users. There are some energy needs which people can live without by postponing their aspirations, and living without the economic and educational benefits that additional energy can allow. However, there are some energy needs which simply cannot be avoided or postponed, such as cooking or lighting. For greater understanding and prioritization, energy needs may be organized into four levels of need: fundamental, basic, productive and, recreation and comfort.

Fundamental energy needs

Fundamental energy needs are associated with human survival and/or extremely poor situations. They are: heating to balance the required temperature of the human body, cooking and lighting. These three energy needs cannot be avoided despite poverty, location or lifestyle.

Heating. People cannot survive if they are unable to keep body temperature to the required level. In rich and cold countries this need is met with space heating using substantial quantities of indoor energy (generally gas and electricity) and with appropriate building materials, and by using appropriate clothing when outdoors. In poor countries, many poor people living in cold regions survive without space heating due to a lack of money or technology. They generally get some warmth in the morning and evening by staying in their kitchen close to the cooking stove, in most cases using the wasted energy from a three-stone stove or other inefficient combustion systems. This often exposes them to the harmful effects of smoke, increasing respiratory infections and child mortality. In hot environments people need to remove heat from their homes, however most poor people cannot afford to do so and have to live in hot conditions.

Cooking. The majority of food has to be cooked before it is eaten. All human cultures cook most or all their food on a daily basis and therefore require some sort of fuel to meet this need. The rich generally have access to liquid and gas fuels, while poor people survive by using a diversity of solid biomass products, wood, agricultural residues, dung and urban solid wastes, or in many cases by spending large amounts on charcoal or wood (the latter mostly in urban poor areas).

Lighting. This is the third fundamental energy need, and again, however poor people may be, some sort of lighting is required during the evening hours for various social, economic or other requirements. These fundamental lighting needs are generally met using very low quality and inefficient energy sources and devices, burning biomass or using kerosene and/or wax candles.

The main concerns with the currently available energy options to meet fundamental energy needs are their low quality and difficult access, in terms of time, inconvenience or cost. The difficulty of accessing these fuels contributes to the 'vicious cycle' of energy poverty and puts a brake on human development. At the same time, the use of these fuels has a negative effect on people's health and on the environment, because of the unsustainable consumption of local resources.

Basic energy needs

The energy required for attaining basic living standards includes all those functions in the previous category (cooking, heating and lighting) and, in addition, energy to provide basic services such as better health, education, communication, transport and others.

These basic services require small quantities of energy but can improve the quality of services greatly. For example, lighting a small health centre may only require a few light bulbs, but it can greatly improve the quality and extend the service into the evening. Lighting can contribute greatly to education by allowing children to study during the evenings, in some cooler regions such as the Himalayas, electricity also lights the classrooms during the day so children can keep warm. For example in the community of Rakel in Nepal, teachers use the electricity from a 200 W wind generator to light the classroom and they keep the windows and doors shut. Also, small cooling boxes for vaccines can make the difference between life and death in isolated communities and only typically require between 400–800 watt-hours per day, but can preserve enough vaccines for hundreds or even thousands of people. The availability of video, TV and electric lighting in schools can, not only enhance the quality of teaching, but also attract more and better teachers. In addition, drinking water in most cases requires small quantities of energy to pump up underground water or to pump from a stream next to the village. Access to information and news is improved with access to media (commonly radios, but to a lesser extent TV and mobile telephones). Finally, family, social and economic links can be enhanced with communication facilities which in turn require a small energy supply.

Energy needs for productive uses and income generation

A range of energy uses are included under this category. These can be grouped as: energy for production, transformation, exploitation of natural resources and for transportation of products to markets.

Productive activities generally require electricity, although some of them can be powered effectively by other sorts of energy, mechanical power or heat. The quantities of energy vary widely, from very small, such as hundreds of watts, to tens of kilowatts.

Energy for recreation and comfort

These needs refer mainly to the use of radio, TV and music hi-fi systems and are associated with entertainment, both at the household level as well as the community level (community or village parties). In addition, convenience appliances, such as washing machines, irons and kettles, may come into the category of 'comfort'. Energy for comfort is generally at the disposal of more well-off families and in urban areas where more facilities and services are available.

At this stage the important questions are: How much energy is the minimum needed to improve the lives of the poor? What sort of energies should be provided to the poor to meet their energy needs, to overcome poverty and to allow them the opportunities of development and sustainability?

The energy needs of the poor

It is very difficult to establish an approximate level of energy consumption required by the poor. This level may depend on a range of factors, such as the physical environment where they live (warm regions will require little or no space heating, while cold regions require more energy to meet this need). As will be seen in further sections of this study, energy usage may depend on:

- the availability, cost and time needed to access existing energy options;
- the sort of activities that the poor engage in to make a living;
- the physical capital that they possess, for example a subsistence farmer may require energy only for cooking and lighting while a farmer with more land may see opportunities to process and add value to some products and to produce for sale in nearby markets;
- the priorities that each individual or family puts on its needs – although lighting, cooking and heating are priorities, other energy uses may depend on household priorities and income levels.

Although there is no dispute that basic energy services contribute to human development and provide the foundation for further developments and wealth, it is not necessarily the case that the more energy supplied the more opportunities the poor have to become wealthy or the greater the increase in the level of the HDI of a certain country, region or village. Experience shows that after meeting basic needs, energy per se is not enough to ensure development. Economic development requires other conditions such as the existence of resources to be processed for sale and physical capital (land, equipment, etc.), which in most cases is not at the disposal of the poor.

Therefore bearing in mind basic energy needs and opportunities, along with the existing evidence on energy consumption, it is essential that the poor have at least energy to meet their basic needs, i.e. cooking, lighting, and sufficient energy to have access to other improved services (such as drinking water, education, health, communications) and opportunities to develop productive activities and businesses.

A variety of sources (Smith, 1994; Barnes et al., 1997; Kaufman et al., 2000; World Bank, 2001) show that energy consumption in rural areas is very small and Practical Action's experience in Peru shows that in isolated rural villages about 60 per cent of the households hardly exceed 20 kWh to 30 kWh per month. By comparison, during the first year of a rural electrification programme in Thailand, the average energy consumption was between 11 and 22 kWh/monthly, and after five years it rose to 22 to 50 kWh/monthly depending on the different villages (Vorvate and Barnes, 2000). Past experience has shown that in rural areas an average of 50 kWh of electricity monthly is what approximately covers the needs of rural families, this is an average per capita electricity requirement of 120 kWh yearly. In addition, the level needed to cover cooking needs is about 35 kg of LPG per capita yearly.

Therefore 'energy access for the poor' should be understood as access to a sufficient quantity of reliable and sustainable energy to serve basic needs (electricity and cooking fuels), and access to good quality devices for cooking.

Energy consumption patterns

Inequalities in energy consumption

To meet these basic energy requirements, i.e. to provide electricity to the more than 1.6 billion presently without this service and to provide LPG for the nearly 3 billion cooking with biomass, would equate to just 1.2 per cent of world total energy consumption. This percentage is derived from estimates for cooking using an average of 35 kg of LPG per capita per year (equivalent to 432 kWh of electricity), which should be sufficient for cooking all meals, and about 120 kWh of electricity per capita per year, which should be sufficient to provide lighting, access to most basic services and to add value to local production.

However, there are currently vast inequalities in both access to energy (sources and services) and energy use not only between developed and developing countries, but also within regions and communities in each developing country.

According to United Nations Development Programme and the World Energy Council (UNDP and WEC, 2004, see Figure 1.4) the average per capita energy consumption – commercial and non-commercial – in OECD North American countries is 11.2 times that of sub-Saharan Africa and Asia, and 6.1 times that of Latin America. These differences may not appear to be very large. The terms 'commercial' and 'non-commercial' hide the fact that energies are used with different levels of efficiencies in different countries, however, and therefore when it comes to useful energy, the differences between OECD and sub-Saharan countries are in fact much greater. In OECD countries people generally use commercial energies with high efficiencies (for example, most of them cook with LPG with efficiencies of 60 per cent or more), whereas the ma-

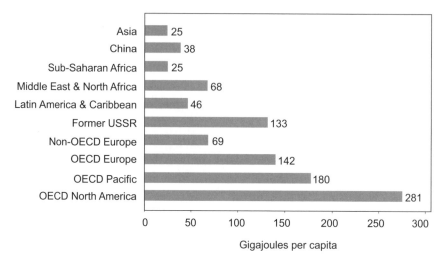

Figure 1.4 Per capita energy use (commercial and non-commercial), by region, 2000
Source: UNDP and WEC, 2004

jority of families in sub-Saharan Africa cook with non-commercial biomass, with very low conversion efficiency (of 10 per cent to 15 per cent).

Great disparities in energy consumption are explained in part by differences in coverage. Table 1.1 gives the numbers and percentages of people in different regions with an electricity supply and it is clear that urban areas fare

Table 1.1 Electricity access in 2005: Regional aggregates

	Population	Urban Population	Population without electricity	Population with electricity	Electrification rate	Urban electrification rate	Rural electrification rate
	Million	*Million*	*Million*	*Million*	*%*	*%*	*%*
Africa	891	343	554	337	37.8	67.9	19.0
North Africa	153	82	7	146	95.5	98.7	91.8
Sub-Saharan Africa	738	261	547	191	25.9	58.3	8.0
Developing Asia	3418	1063	930	2488	72.8	86.4	65.1
China and East Asia	1951	772	224	1728	88.5	94.9	84.0
South Asia	1467	291	706	760	51.8	69.7	44.7
Latin America	449	338	45	404	90.0	98.0	65.6
Middle East	186	121	41	145	78.1	86.7	61.8
Developing countries	4943	1866	1569	3374	68.3	85.2	56.4
Transition economics and OECD	1510	1090	8	1501	99.5	100.0	98.1
World	6452	2956	1577	4875	75.6	90.4	61.7

Source: IEA, 2006

better than rural areas, and North Africa is well supplied compared to sub-Saharan Africa.

There is an increasing consumption of liquid fuels in developing countries as urbanization increases, with the shift from one fuel to another (generally the most convenient in terms of comfort and efficiency of conversion) depending on the increase of income by the population. An example of fuel switching with economic development is Korea, where cooking is now based on liquid fuels. In China, electricity for household use grew four fold in one decade (from 1980 to 1990) from 3 to 13.6 million ton oil equivalent (TOE); in the same period use of gas also increased from 1.2 to 4.6 million TOE.

Regarding electricity, World Bank statistics show that average per capita consumption in the higher income countries is 25 times the average in South Asian countries. However, comparing large regions using average figures hides extreme difference between poorest and richest countries. For example, per capita electricity consumption in the USA in 2001 was 13,052 kWh and 6,000 kWh in the European Union, while in Vietnam annual per capita electricity consumption was 331 kWh, 116 kWh in Kenya, 275 kWh in India and 127 kWh in Yemen. Furthermore, while rich countries consistently increase their yearly electricity consumption, poor countries experience little or no improvement except for a handful of fast-growing economies, such as China, Brazil, India and Mexico.

Inequalities are also present within the different population strata in developing countries. In capital cities and large towns, per capita electricity consumption is several times higher than that in small towns and villages with electricity access. In large cities and towns there are projects and continuous investment to cope with, future demands and energy security is always a concern; while for rural villages, in most countries, private investors are not interested and public investment is simply absent.

While average domestic monthly electricity consumption per household in urban regions of developing countries is around 50 kWh to 100 kWh (for those with access to electricity), there is evidence that in rural areas of developing countries the energy consumption hardly exceeds 30 kWh to 50 kWh per month (CER-UNI, 1998; Smith, 1994; UNDP, 2001). Practical Action's experience in several countries (Peru, Nicaragua, Nepal, Sri Lanka and others) shows that in isolated rural areas about 60 per cent of the population hardly use more than 20 kWh. Other papers show consumption figures even smaller than these, for example, impact studies show that during the first year of the rural electrification programme in Thailand the average household energy consumption was between 11 and 22 kWh/monthly, and after five years it had risen to 22 to 50 kWh/monthly depending on the village, despite the fact that rural electrification in Thailand was mainly based on grid connections (Sanchez, 2006b). According to Goldemburg et al. (2004) household electricity consumption has increased in every continent, but mostly in regions or developing countries with higher growth rates (for example, China, India, Mexico, Brazil and others).

Besides the small amounts of energy (per capita) required, another important characteristic of rural electricity demand is the dispersion of consumers, as most rural populations live in small dispersed villages, communities or settlements, making it difficult and expensive to access commercial energy services, among them electricity from national grids.

Inequities extend to the whole range of services and sectors that depend on energy, including transport. For example, while per capita annual consumption of energy in the transport sector in OECD countries ranges from 16,000 to 22,000 kWh, in low-income countries this ranges from a few tens of kWh to as much as 2,000 kWh (Cozzi, 2006).

Poverty and a lack of financial mechanisms prevent the poor accessing modern fuels for cooking and heating. To give an idea of the order of these costs, a very small gas stove together with a 12 kg LPG bottle would cost around $100, a 50 W solar PV system in an isolated rural area would cost around $800, and a micro or mini-hydro plant would cost over $2,000 for each kW installed.

Isolated rural communities cannot afford electricity because they cannot pay these upfront costs and because the grid is simply out of their reach. In addition they face difficulties buying fossil fuels not only because they cannot pay, but also because the supply is generally unreliable due to transportation limitations. Therefore most of the time the poorest and most marginalized end up excluded from the benefits of the subsidies typically used in developing countries, such as subsidized electricity tariffs, subsidized cooking fuels (kerosene and LPG) or subsidized public transport sectors.

A lack of knowledge and information on energy benefits and options, prevents the poor from voicing their concerns about energy, therefore they are often ignored or excluded from energy planning or policy making. This lack of information makes the poor a market that is unaware of energy issues, of the constraints to implementing energy systems, of the existence of energy options and costs – therefore a difficult market to deal with when it comes to paying for energy used. In Peru, for example, it was apparent that people often think that once a small hydro system is installed it can produce energy for free; therefore they don't see the need to pay for it. It takes time and resources to actually explain the facts, including the realities behind energy systems and their cost. Eventually when these issues are understood, however, people generally assume responsibility for paying for running costs and looking after their energy systems.

Unequal access to energy options

Basic energy needs (cooking, heating and lighting) are similar for rural and urban households, however, the supply sources and costs are different. While most rural people can easily access agricultural residues, dung and wood for cooking for free (other than the cost of the time taken to collect it), urban people mostly purchase fuels (traditional biofuels as well as LPG and kerosene) from the market.

The strategy used by the poor to cope with their energy needs is generally to go for the cheapest possible option; however, that strategy does not always work with the poorest. In many cases the poorest people are deprived of using the lowest cost energy option, such as subsidized LPG or kerosene simply because they cannot purchase the cookers needed. A study carried out by Practical Action in Al Fasher in Sudan (Hood, 2006) showed for example that people keep using wood and charcoal despite the fact that they cost about three times more than LPG because they are unable to purchase a gas cooker.

Statistics show that in urban areas there is much better coverage of electricity supply, hence urban people are more likely to use it for lighting, however in many cases costs may be the main constraint. In fact there are many examples where people are connected to electricity but cannot afford to pay the bill, and are therefore soon disconnected from the electricity service. This also happens in rural areas where there is electrification. In rural areas without access to electricity people use a range of energy sources, such as kerosene lamps, wax candles, and dry and wet batteries (dry batteries are used mainly for torches, and small radios, while wet batteries are mostly used for lighting, small radios and black and white televisions).

A World Bank (2003a) study on fuel switching in eight countries in Africa, Asia and Latin America confirms that there are differences in fuel consumption in urban and rural areas. It shows that solid fuels are used in much lower proportions in urban areas than in rural areas where the study had been done. It also shows that the use of solid fuels in urban areas declines as family income increases, while solid fuels are widely used in rural areas even within the top expenditure brackets. The study shows that urban people expend a larger share of their budget on cooking fuels than those in rural areas, and that among all those who purchase fuels, firewood has the highest budget share among its users (energy accounts for about 5 per cent of a household budget as an average for all users, and about 10 per cent to 15 per cent for those using only wood for cooking). The different percentages are influenced by differences in accessibility, and transport costs are affected by accessibility: in large towns and cities the fuel costs are cheaper than in small isolated towns with poor vehicles and rough roads.

The study concludes that 'large groups of households particularly in rural areas of low-income countries will remain unrealistic targets for fuel switching for quite some time to come – other interventions may be more appropriate for such households, such as improved stoves, better ventilation – keeping in mind the minimum purchase power of this group, it will require that effective technologies are available at low cost'.

Unequal access to energy in large cities in Latin America

A World Energy Council (WEC, 2006) case study on energy access for the poor in three large Latin America Cities, Rio de Janeiro, Buenos Aires and Caracas, showed that physical access to energy is not a problem in any of these cities, since all three are close to 100 per cent connected to electricity, and all have access to cooking liquid or gas fuels. The main problem is cost. The portion of the household budget spent on energy in poor households was reported to be an average of 7.6 per cent for Buenos Aires, 3.3 per cent for Caracas and 15.6 per cent for Rio de Janeiro, while the budget share for poor households in the UK is an average of 4.1 per cent and in the USA 5.2 per cent. The researchers also recognize that the lower cost in Caracas is due to heavy subsidies, in Buenos Aires some subsidies, while in Rio there were no such subsidies. These results clearly reveal that for these cities, energy access for the poor depends upon the capacity to pay, but they also reveal that there are serious inequities in the benefits from subsidies (in Caracas, for example, those who benefited most from the subsidies were not the lower-income sectors). This study also revealed that there is a high percentage of illegal connections in all three cities.

The poor miss out on energy access in Zambia

Further evidence from Zambia confirms that subsidies are often captured by the non-poor. Kalumiana (2004) in his article on energy services for the poor (executive summary) shows that:

> In Zambia at least 56% of urban residents are poor. Only 48% of the urban population has access to electricity. Urban residents in Zambia depend on four main fuels: electricity, charcoal, firewood and kerosene. Kerosene is mainly used for illumination, electricity for both illumination and cooking while charcoal and firewood are used mainly for heat-related end-uses – mainly cooking. Electricity is the most important lighting fuel for urban residents. Of these four fuels, electricity is subsidized...The bulk of the total energy subsidies arising from pricing of electricity is captured by the non-poor. In 1998, the non-poor captured Ks8.8 billion ($4.2 million) in electricity subsidy, which represented 83% of the total subsidy on electricity.

This implies that while rich people in Zambia have benefited from electricity subsidies (including for cooking) poor people are excluded from such subsidies.

Energy to be required and used for productive purposes will depend largely on market opportunities arising and productive activities developing. Much has been said about energy for productive uses and the importance of this demand for energy in boosting the sustainability of energy services, as well as the potential of these income opportunities to help people escape the poverty trap. It is less clear, however, how to use energy most effectively to make marketable products in different situations.

There are important differences between the productive energy needs of the poor in urban areas and energy needs in rural areas. In towns the energy needs are associated with running very small shops and with the transformation of products, while in rural areas energy is needed for a wider range of applications.

Rural inhabitants need energy for farming (including irrigation, land preparation, crop cultivation and so on), animal husbandry, transporting products to markets, processing products and value addition (milling, carpentry, cooling food for preservation etc). They also need energy for extraction and the transformation of local natural resources, such as natural products related to small mining or building materials (like ceramics, roofing products, extraction of sand for construction and more). A recent study in Peru (Sanchez, 2006b), shows that from 5 to 7 per cent of families living in small villages with access to electricity run small businesses or engage in service provision, although those small businesses and services are for the local market rather than for selling in larger markets.

Summary

- About 1.62 billion people worldwide have no access to electricity services; nearly 3 billion people lack modern fuels for cooking.
- To make progress in development, countries need access to sufficient energy for everyone. There is a correlation between energy consumption per capita and the Human Development Index (HDI).
- To deliver basic energy requirements for the two billion people currently without access to modern energy would equate to just 1.2 per cent of the world's total energy consumption.
- The energy needs of the poor can be classified as: fundamental, basic, relating to production, and to comfort and recreation.
- The potential energy options for rural and urban people are different. While for rural people small decentralized energy systems are more appropriate to provide electricity, for urban customers the grid is better.
- For cooking, biomass will continue to be the main fuel for rural people, while liquid fuels (kerosene, natural gas and LPG) are reaching more and more people in urban areas.
- Subsidies, if not properly managed and targeted, benefit the better off much more than the poor.
- In many cases the poorest cannot access the cheapest and/or most convenient option because they cannot afford the cost of appliances, such as cookers, or efficient bulbs.

Notes

1. An 'improved energy source' for cooking is one which requires less than four person hours per week per household to collect fuel and meets the recommendations of the WHO (World Health Organisation) for air quality (maximum concentration of CO_2 of 30 mg/m^3 for a 24 hour period and less than 10 mg/m^3 for a period of 8 hours of exposure). Plus the overall conversion efficiency must be higher than 25 per cent.

CHAPTER 2
Failure to reach the energy poor

Why has so little been achieved to date?

The big question is why the modest amounts of energy needed are not yet available to the poor. The progress on access to electricity by the poor has been low and inconsistent during the last 40 years; in 1970 the population without access to electricity was 1.9 billion, in 1990 it was 2 billion (Simões, 2006), and the present statistics show that about 1.6 billion have no access (IEA, 2006).

The answer is found by looking at the economic situation of most people in poor countries. The great majority of people without access to energy live on less than $2 per day, making it difficult to access good services, including access to modern energy sources. Energy access is not cheap; the initial investment to set up generation systems or to arrange for a connection (electricity) or to purchase better devices (in the case of cooking) can be high, and if it has to be paid for by the poor, they will generally struggle to do so. Even after the initial access is established, a poor rural family may prioritize the purchase of medicine for a sick child or basic food ingredients rather than pay the electricity bill.

People in rural areas generally cook with solid biomass which is commonly collected free of cost – though it can take considerable time to collect. Most of these people would therefore find it hard to pay for liquid or gas fuels even if these were available. The reality is that rural people generally do not have many income-earning opportunities to exploit. Developing country governments have little capacity to fund many of the competing needs of the poor and energy access is often neglected; hence the huge financial gap.

Before we consider the barriers to energy services and the government reforms and international trade policies that have affected progress, it is useful to examine how developed countries managed to achieve rural electrification in the early twentieth century.

Lessons from developed countries' rural electrification

Since the late nineteenth century, when the era of electricity started, the extension of energy services to people has varied from country to country. Developed countries have used energy extensively to industrialize and thus to increase their wealth. They accessed modern energy in a relatively short period and progressively improved technology and reduced costs. However, even in rich countries, access to energy and especially to electricity has not been straightforward. These countries had to implement imaginative strategies and provide extensive subsidies to stimulate electricity use and markets. The cases of USA and Europe are of particular interest due to their early development of energy usage.

According to Lapping (2005), by 1920 about two-thirds of rural farmers in Germany, France, the Netherlands and Scandinavia had access to electricity; while in the USA 90 per cent of farmers had no access. The European farmers had very favourable rates for electricity (subsidized by government) that allowed them to use energy in almost all activities. At that time in the USA private entrepreneurs were reluctant to invest in rural electrification because of the high cost of infrastructure.

PowerSouth Energy Cooperative is the generation and transmission cooperative (USA) that provides the wholesale power needs of 20 distribution members – 16 electric cooperatives and 4 municipal electric systems – in Alabama and north-west Florida. They wrote (PowerSouth Energy Cooperative, 2009):

> Because there was no electricity, life and work for most rural Americans in the 1930s was fixed in a cycle of hardship and drudgery, little had changed from decades before. Only the most affluent farmers and ranchers or those near towns could get 'the electric'. The majority of rural people, nearly 90 per cent, lived and laboured in a dark and powerless land. But by the beginning of 1935, the hope for electricity in rural America became a reality. The federal government had a plan to provide assistance to the people of rural America and help them bring the power into their lives. They could organize themselves cooperatively and secure loans to electrify in the rural areas. This message swept the land. It was the beginning of the full-scale electrification of America, a partnership of the people with their government. People called it 'the REA' (Rural Electrification Administration).

As is now the case in most developing countries, rural electrification in the USA during the 1930s was not an attractive business for private investors. Due to the dispersion of the rural population, the cost of electricity in rural areas in the 1920s was double that of the city. During the 1920s there were important debates about the way forward regarding electricity access in the USA; people like President Herbert Hoover argued that the responsibility for rural electrification rested with state government.

In 1935 the US federal government created the REA and provided funds for its operation; thus the cooperative model for rural electrification was adopted. The REA was essentially a government-financing agency providing subsidized loans to private companies, public agencies, or cooperatives for the construction of electrical supply infrastructure in rural regions. The loans were guaranteed by the federal government and had an attractive interest rate and a generous repayment schedule of 25 years. The REA extended its rural electrification programme by periods of ten years until the end of the twentieth century. By 1998 the rural electrification cooperatives provided electricity to 11 per cent of the US population.

As mentioned earlier, the rural electrification process in Europe also started with a combination of incentives, subsidies and government investment. In France rural electrification was implemented with the participation of local government and people, but also with central government decisions and investment.

In summary, rural electrification in developed countries was successful thanks to government willingness to invest and to clear policies regarding subsidies for investment and tariffs. Developed countries have achieved access for all their citizens and now the problem is no longer investment, it is efficiency, productivity, opportunity, and so on.

An important question that arises in discussions among rural electrification promoters in developing countries is: why are the successful experiences of the USA and Europe not emulated by developing countries, and why haven't their strategies been adopted? In fact there have been many attempts to emulate these cases, especially adopting the US cooperative model, but with few exceptions such attempts have failed dramatically. There are clearly barriers to the success of rural electrification in developing countries, despite the use of previously successful delivery models or approaches.

Before discussing these barriers, it is important to point out that the failure to emulate previous successful models may be due to the fact that energy promoters in developing countries do not have conditions such as a strong economy, high productivity, good markets and other conditions, which were prevalent in Europe and the USA at the time of rural electrification. For example, the average per capita income in the USA was about $15,000 (equivalent to the purchasing power of $15,000 in 2003) by 1960, and rural farmers were not the poorest people in the USA, hence electricity at that time was required for economic production and growth as well as for social development. The socioeconomic conditions of the developing countries at present are different: in most African economies per capita income hardly reaches $1,000, whereas in Latin America with few exceptions per capita income is below $3,000.

Chapter four gives examples of two developing countries – Peru and Sri Lanka – which are struggling to supply electricity to the rural poor. While policies are in place for electricity expansion, they are unable to overcome the barriers of remoteness and poverty.

Is the issue of sustainability misleading?

For the purpose of this book, the definition adopted for sustainable energy is that adopted by the Sustainable Energy Group (2008): 'sustainable energy means meeting our needs today without compromising the ability of future generations to meet their needs'.

The issue of sustainability

Sustainability of energy access for the poor – that is the continuous supply of reliable energy services long after any initial energy access programme has ended – has been a great concern, especially from the early 1980s onwards. In urban areas, in some cases after only a few months of connection to electricity, poor people stop using it; or despite the regular supply of kerosene, natural gas and LPG, people continue using wood, charcoal or biomass residues. In rural areas, small energy generation systems installed to provide electricity to small villages or communities frequently last only a few months before they are abandoned. Similarly, large numbers of projects disseminating fuel-efficient stoves have not changed the common practice of using three stones. These situations can be seen as 'unsustainable energy access for the poor', and are caused by the original programme not addressing underlying barriers associated with poverty and household cash flow, lack of technical capacity and institutional support, and with social and cultural issues.

Small-scale standalone energy generation systems such as micro-hydro, diesel sets, solar PV (photo-voltaics), small wind and biomass energy generation systems are considered viable technical and financial solutions to meet rural energy needs. However, the failures reported in many parts of the world have created a lack of confidence on the part of government and planners regarding the sustainability of these systems.

Over the last three decades, universities, research institutions, NGOs and other organizations have undertaken a large number of research programmes and piloting activities in order to prove the technical and financial viability and the sustainability of small standalone energy generation systems. Governments in developing countries, multilateral and bilateral organizations (e.g. World Bank, UNDP) have been searching for strategies to scale-up such systems. However, most of the literature reports little success in the scaling-up process and the few successful cases, such as China (see page 62), have happened with considerable government support, including subsidies for investment and in many cases for operation as well.

The long drive for sustainability

The failure of these pilot schemes or dissemination projects has created an atmosphere of urgency for achieving sustainable systems for energy access for the poor. The academic world, NGOs, multilateral and bilateral organizations have attempted to tackle the problem with pilot projects and with policies and strategies. Their attempts have followed two main fronts of action: policies and strategies; and technologies and approaches. However, all actors have adopted common concepts that have changed with time mainly as the previous concept has dropped out of favour due to a lack of success. These concepts are linked to certain myths which are described at the end of this chapter. The failure of the electricity reforms which were based on these concepts are described in the following section.

Concept one: Maximize energy use

During the 1980s, the dominant concept aimed at increasing sustainability was economic development. Under this concept energy had to be a 'development vector' and had to be for income and growth. Ideas such as 'rational use of energy' or 'high load factor' were considered by development organizations to be key to success and sustainability. In other words, the more energy used, the higher the benefits and the greater the possibility of sustainability.

This concept encouraged research and development of some specific technologies to make maximum use of available energy. Among the research topics were energy storage and new end uses for surplus energy and for increasing production.

For example, in the 1980s and early 1990s the UK Department for International Development (DFID) funded two projects of this kind associated with cooking with surplus power from micro-hydro projects:

- The Bijuli Dekchi (BD) – a low-wattage cooker which makes use of off-peak electricity to heat water for cooking. The hot water is especially useful in cooking rice and lentils. This project was developed and tested in Nepal, where a few units have been built and used, but the results have not helped to disseminate or sustain existing micro-hydropower systems.
- The Air Heat Storage Cooker (HSC) – an electric cooking stove which stores off-peak electricity by heating pebbles. The energy is released by blowing air over hot pebbles and the hot air then heats the cooking pots. Since the temperature of the air is high it can be used for frying.

Several projects on energy for productive uses were promoted and implemented during the 1980s and 1990s. These projects have had two components, the energy generation system itself and the productive part, most related to the processing of local products. Such programmes achieved only limited success; in most cases the productive activities failed before the energy generation scheme. Among the reasons for the failure of the projects were that small, poor communities generally have neither access to markets, nor capital, or the necessary training. In addition the production of products for commercial markets requires management skills and entrepreneurial knowledge, as well as a permanent supply of raw materials and resources, an area the projects or programmes did not pay attention to.

Concept two: Pay for service

During the 1990s, the approach promoted by development agencies, led by the big financial agencies (such as the World Bank and others) was 'maximum cost recovery'. If all energy users paid as much as possible for the cost of their services this would promote a sustainable market in energy service (Barnes and Foley, 2004; ADB, 2005). It would also gradually reduce the need for subsidies which are a burden on public finances. The emphasis on cost recovery was also a response to the belief that payment by the poor would create a feeling of ownership and ensure responsibility towards and care for the system. 'Prices and fee structures should be low enough to attract customers but high enough to cover capital, financing, servicing, equipment replacement, and administrative costs, as well as possible defaults.' (Cabraal et al., 1996) The goal of full cost recovery has frequently been confronted with the reality of the lack of capital on the part of the poor and therefore credit systems have been promoted to tackle this problem. This concept has inspired a range of imaginative delivery models for energy, among them the 'fee for service' model, which has been largely promoted by the World Bank and some other development organizations. The projects of SOLUZ in Central America and the Caribbean Islands are among them. Unfortunately projects based on this concept turned out to be unviable for most of the poor, because the majority do not have sufficient income, especially when the attempts to cover costs include the initial investment.

This issue was promoted during the period of deregulation and privatization of state properties. Most governments therefore adopted the line that full recovery might be possible and consequently focused on privatization without thinking about energy access for the poor as a problem, thus leaving all in the hands of market forces.

Concept three: Energy to meet all development needs

From the early 2000s to the present day, the main concept aimed at ensuring sustainability has been the partnerships and integrated approaches that include all sectors (energy, but also health, education, agriculture, communication, development, etc.) and encompass all the needs of the poor. The aim has been to investigate and foster conditions for income generation and hence increase capacity to pay for energy services. This approach also considers it essential to have full agreement with all parties involved (civil society, government, users, entrepreneurs, donors etc.), and that only by including all the needs of the poor will projects become sustainable.

With this approach came the idea of partnerships to tackle sustainability. These partnerships include GVEP (Global Village Energy Partnership), REEEP (Renewable Energy and Energy Efficiency Partnership) and GNESD (Global Network on Energy for Sustainable Development). These partnerships work in different areas and with different approaches. In 2004 and 2005 GVEP was initially mainly dedicated to advocating for energy access for the poor with a focus on this multi-sectoral approach; currently it is an international NGO with a focus on service delivery. Both REEEP and GNESD focus on policy research and advocacy. There has been some criticism that these partnerships have not lived up to their expected function of improving the focus on energy access and poverty reduction. A common weakness of these partnerships is that a majority of members are from the North rather than from the South, which indicates that Southern members are either not enthusiastic about this approach as a solution or that they have not been consulted properly about the potential opportunities.

So far this concept has demonstrated little success. The integrated approach is generally at odds with common country policies and practices, because at this level all activities are handled by sectors, and each sector adopts its own priorities and strategies. The question here is whether such an inclusive and broad approach can be effective? Unfortunately, there has not been previous experience in any developed country that can cast light on this question. In fact, a simple way of looking at this concept is that it attempts to solve the whole problem of development instead of energy access only. Furthermore, this approach is hardly compatible with private investment. It is difficult to imagine private investors carrying out a wide consultation before investing their money in energy services and the fact remains that private investment will always require complementary investments in one form or another from governments and international donors.

Global trade policies

These have also had an effect on the drive for sustainability. Globalization trends have become increasingly important for the world economy; this has meant the creation of new trade policies and mechanisms. Trade in services have been one of the most important issues for international businesses.

GATS (General Agreement on Trade in Services), is the most important mechanism to accelerate the growth of international business in the field of services and is a legally binding mechanism between signing partners. Its roots go back to the 1947 General Agreement on Trade and Tariffs (GATT), which was the first major international agreement aimed at reducing barriers, such as tariffs and dumping, to free trade. The Uruguay Round saw, for the first time, a major focus on trade in commodities other than goods, for example, services and intellectual property. The Uruguay Round culminated, in 1994, with the signing of the General Agreement on Trade in Services (GATS), which came into force in January 1995.

A new round of negotiations was launched in 2001 at Doha under the banner the 'Doha Development Agenda', part of which was the GATS 2000 initiative. The aim of the Doha Round was to promote development by lowering trade barriers to trade in agriculture, industrial products and services. Other issues that form part of the Doha Round include negotiations on trade-related intellectual property rights and special and preferential treatment for the least-developed countries. Several ministerial meetings and round tables have been held since then. The last series of meetings related to GATS was held in July 2008 in Geneva.

The GATS agreement has two parts. In the first, the general rules are set out, while in the second there are national 'schedules', which list individual countries' specific commitments on access to their domestic markets by foreign suppliers. Subsequent rounds of negotiation have concentrated on increasing the number of national service markets open to competition. These have produced 'protocols' with further commitments on financial services, basic telecommunication services and movement of 'natural persons'. The WTO acknowledges that in the latter case little has been achieved.

The main concern regarding GATS is that advocating universal privatization and the opening up of services to international commercial interests, relies on market mechanisms as the driving force for the provision of services. This, however, is not the best way to reach the poor. So far there are a large number of disputes between communities and providers, and between government and international companies about the accomplishment of agreements.

Moreover, while the World Bank and other organizations promote the participation of the community, inclusiveness and other principles as the way forward for the provision of services, the WTO promotes secrecy in negotiations. The secrecy of the GATS negotiations leads to a lack of participation from different parts of civil society in decision making.

In addition, the process of GATS is being negotiated under unfavourable conditions for the governments of the south where there is an unequal balance of forces between southern and northern governments in these negotiations. GATS does not consider the social and economic environment of the communities receiving the service provision, or their governance schemes and cultural legacy.

In most countries there has been little trust towards the government as a service provider and much of the disenchantment is due to government corruption, yet GATS does not require any improvements in governance as a prerequisite. The negotiations take place with the same political class and with the same socio-economic elite of the countries who previously managed these services as government appointees. Transparent and honest governance should be one of the main issues before GATS if success is to be expected. Privatization in many countries has been done before establishing proper regulations, and this, together with corruption and poor governance, has been favourable to international companies which, in the absence of virtually any rules, have been able to organize their business in their own best interests (see Collier, 2007).

The failure of the electricity reforms to provide energy access for the poor

Electricity reforms in the 1980s and 1990s were promoted mostly to reverse the chaotic situation and poor financial performance of state-owned utilities in developing countries. They were seen as key to attracting foreign investment and extending the services, but they were also considered key to accelerating people's access to electricity services. The electricity reforms in developing countries started in Latin America: in Chile in the early 1980s and then to most other countries in this continent in the 1990s. In Africa and Asia reform has been moving slower since the early 1990s and in many cases it is still to start. For those countries that have finished or are in the late stages of reform, the lessons are that reforms brought positive impact in terms of investment and efficiency in the sector (less technical and non-technical losses), more generation capacity available and more security of energy supply. However, there is evidence of negative impacts on energy access for the poor.

A report from the GNESD about the impact of electricity reforms in Africa, Asia and Latin America (2004) reveals that urban electrification has either increased by a small amount, stayed level or slightly decreased, while rural electrification rates have severely decreased as a consequence of the reforms. For example, in Kenya the rural electrification rate changed from 16.1 per cent pre-reform to 7.7 per cent post-reform, and for Senegal from 16.1 to 1.4 per cent. The study states 'market-oriented reforms have either had a neutral or adverse impact on the poor and should be redesigned especially if the reforms are to be justified under the poverty reduction agenda. The key identified negative impacts on the poor include: reduction in electrification rates, increased tariff levels and a decline in electricity consumption' (GNESD, 2004: 32).

Powell and Starks (2000), writing for the World Bank, consider that electricity reform is based on the premise that market mechanisms supply electricity much more efficiently than central planning. However, they clearly express their concerns regarding the effectiveness of such reforms on energy access for the poor. In their introductory paragraph they state: 'unless energy can be produced and delivered more cheaply, it will stay beyond the reach of the poor. For energy delivered through networks the costs that matter are not only the unit costs but the costs of extending the networks' (Powell and Starks, 2000: 1). They argue that electricity reforms have contributed importantly to cost reduction, but not enough to make electricity available for the majority of the poor, and that geography plays a role in the distribution costs as well as supply costs. Furthermore, they argue that it has not been possible to pass the reduction in costs in energy generation (which comprises the highest share of the costs in the electricity service chain) effectively to the poor and that tariff policy has to be designed with their needs in mind. They also state that transmission and distribution costs have not fallen sufficiently to extend the grid to all areas. Finally they conclude that, as long as the reforms encourage private competition and profit-seeking combined with regulation and tariffs sensitive to the needs of the poor, the reforms are positive.

Williams and Ghanadan (2006) made a comprehensive analysis of non-OECD countries and argue that reforms should be based on a realistic assessment of a country's needs and capabilities. They argue that the reforms in non-OECD countries have followed a standard menu that has been based on the success of the OECD electricity reforms, and that such a menu has not taken into consideration the differences between the situation of OECD and non-OECD countries at a time when the respective reforms happened. While in OECD countries reforms were driven by the financial situation of their macro economies, which were socially and politically stable, the reforms in the non-OECD countries have happened under unstable social and political conditions.

Examples from Bolivia, Ghana, India, Poland and Thailand, with different backgrounds and geography, are given as evidence. For Bolivia they conclude that 'reforms will be seen as a technical success but a social and political failure'. In Ghana, after ten years of reform the sector structure remains much the same as at the beginning, especially because the foreign private company that invested in the reform revoked the contract and left the country due to social and political disruptions. In India electricity reform remains a contentious political issue, especially after the 2004 elections were lost by the party involved in the electricity reforms. In addition, three states have recently offered free electricity to farmers. In Poland, the reforms created a huge burden on the economy of the poorest sectors, due to the continuous adjustments required to meet production and distribution costs. At present the World Bank (which led the reforms initially and then progressively left as Poland gained access money from the EU) considers that the Polish reforms have failed due to the lack of social pricing polices. In Thailand, although the state-owned utilities performed well on electrification coverage (90 per cent of rural villages), by early 1990 the sector had become a great burden, with debts and needs for further investment. Therefore, although reforms for private participation have been promoted ever since, due to internal political reasons, no progress has been made on this front.

The World Bank has been the most important driver of all these country reforms and the leading policy designer and money lender.

The barriers to rural energy supply

We have seen how successive energy reforms based on a number of concepts have made little progress towards energy access for all. A number of studies have analysed the barriers to energy access for the poor and to the sustainability of the systems. Table 2.1 summarizes the findings of a number of these studies.

In summary, the barriers to energy access are as follows:

Rural energy barriers. There are a number of barriers to rural electrification or improved rural energy supply which have not yet been overcome despite the efforts of many organizations and researchers. These barriers relate to local, national or regional contexts:

- poverty of potential customers means that it is difficult to attract private investment and overstretched budgets at the national level mean that energy access for the poor is a low priority;
- investment costs for grid extension are high due to the low density of the population and the high investment costs of small decentralized energy generation;

Table 2.1 Studies into barriers to the poor accessing modern energy

Research study	Barriers to the poor accessing modern energy
Koharznovich (2008)	Decentralized renewable energies confront barriers on: • limited information; • lack of technical skills; • inaccessibility to technology; • lack of institutional capacity; • prohibitive cost. Proposed solutions: • people centred solutions, where communities have a voice in decision making, and all the sectors are included; • should work at all levels, and be holistic; • bottom-up management and regulation of the sector; • increasing the choice of energy services.
Beck and Martinot (2004)	Policies must address: • renewable energy promotion – investment costs, pricing, risks, transaction costs, interconnection requirements; • biofuels – lack of production and high cost, emissions reduction policies, tackling environmental externalities; • power sector restructuring – monopolies, subsidies, unbundling electricity sector and other market mechanisms; • distributed generation – costs, pricing and interconnection barriers; • rural electrification policies – lack of subsidies, lack of skills and lack of credits.
Painuly and Fenhann (2002)	Barriers for the implementation of renewable technologies for rural electrification: • institutional (relating to research and development, and demonstration and implementation); • market (small size of the market, limited access to the international market, limited involvement of private sector); • awareness of information (lack of access to information on renewable energy technologies (RETs); • financial (inadequate financial arrangements, local, national and international for RETs projects); • economic (unaffordable costs, taxes on imports, subsidies and energy prices); • technical (lack of access to technology, inadequate maintenance, low quality of products); • capacity (lack of skilled manpower and training facilities); • social (lack of social acceptance and local participation); • environmental (visual pollution, lack of valuation of social and environmental benefits); • policy (unfavourable energy policies and unwieldy regulatory mechanisms).
Sanchez (2006b)	Electricity access for the poor in rural areas confronts important barriers related to technological, economic, financial, social, and policy issues. In the local context the critical factors for success are: • local technical capacity; • local technical skills to operate, administrate and maintain isolated energy schemes; • appropriate management models. At the national level, critical factors for success are: • national capacity to manufacture equipment and spare parts and provide technical assistance;

ESMAP
(2000)

- regulations;
- access to markets and information.

'Moving from traditional biomass to modern fuels can thus dramatically raise the effective income of low-income households. But substantial barriers may prevent low-income house-holds and communities from gaining access to modern energy services'. These include:
- high connection costs to networks;
- high cost of electric appliances;
- low density of demand;
- high investment cost of small decentralized energy systems;
- lack of commercial mechanisms to deliver products and services between suppliers and customers;
- mechanisms ill suited to dealing with poor customers in informal settlements (which could have, for example, informal addresses);
- lack of credit for the poor;
- customers having difficulties paying the bill in those countries where previously they relied on highly subsidized services (Eastern Europe and Central Asia).

This same study argues that sometimes government policies inadvertently limit the access of the poor to energy; among them:
- long-term franchise arrangements may block the opportunities for alternative energy services supply (because of the monopoly);
- tariffs, taxes and subsidies sometimes can inhibit grid extension in rural areas;
- over specification of technical standards and quality increases costs unnecessarily.

- lack of national capacity in developing countries to produce energy generation equipment and spare parts, hence high costs of importation and inappropriate implementation;
- lack of financial mechanisms to enable people's access to credits to cover costs of implementation, connection, spare parts, repairs and services;
- lack of appropriate legal frameworks to promote decentralized energy schemes, community management models;
- difficulties in applying effective subsidies for the poor.

Barriers to access to fuels for cooking in rural areas. Rural people face similar barriers to electricity when trying to gain access to modern fuel for cooking (LPG, kerosene, electricity):

- dispersed population, hence high cost of distribution;
- lack of capital for the purchase of appliances when modern fuels reach their villages or communities;
- lack of appropriate incentives and policies for afforestation and forest management, hence scarcity of fuels, and long hours spent collecting biomass.

Energy access for the urban poor. In urban areas access to electricity, as well as access to cooking fuels are problematic. Here the supply is generally available in the market or could easily become available, however the market mechanisms cannot reach those in poverty. The most important barriers are:

- low income of the poor, hence no capacity to pay the full cost of energy;
- poor policies on subsidies, which favour the middle classes above the poor;
- a lack of financial mechanisms to facilitate the purchase of appliances and connection to grids (electricity) or suppliers (gas sellers);
- living in informal settlements without formal ownership of the land or house.

Barriers to access for the poor – myths or realities?

As we have seen, over the last 20 years or so there has been extensive research into understanding the barriers and solutions to energy access for the poor. As a consequence, some important barriers much discussed in the past are now considered by some to be myths. However, these opinions are controversial.

DFID refers to the existence of five myths (DFID, 2002) on energy access by the poor and therefore argue that such myths need to be dispelled in order to make progress in energy access for the poor. These myths are: 1) poor people do not consider energy to be a priority; 2) electricity, whether from grid or from decentralized energy sources, will solve all the energy service needs of the poor; 3) the poor cannot pay for energy services; 4) new technology alone, such as solar photovoltaic cells, will improve people's access to energy services; 5) only people in rural areas suffer from a lack of energy access.

Saghir (2005) of the World Bank states:

Myths about energy and poverty abound. Among the most pervasive is this: for poor people who use biomass, that energy is free and they are insensitive to changes in energy prices. Another myth is this: when the poor have to pay for their energy, that energy is cheap compared with the modern energy used by the wealthier households. And yet another: when modern energy is first introduced in an area, it is cheap and easy availability will prove a panacea, kick-starting enormous socio-economic development.

UNDP energy expert Susan McDade debunks four myths on energy for the poor (Columbia Business School Alumni, 2004). She believes that it is a fallacy to think that 'energy is not a priority', 'poor cannot pay for energy services', 'electricity will solve the energy crisis' and 'sustainable development does not include fossil fuels'.

The reality is that some of these may be myths, but some of them may be crude realities that we may not want to accept. Some myths should be overturned and there is abundant literature and experience of many field workers to refute the notion that:

- people do not see energy as a priority;

- 'new technologies alone will improve people's access to energy services' – rather there are a number of institutional and social constraints that are more important limitations than technology alone;
- 'electricity will solve all the energy service needs of the poor' – instead the poor need a range of fuels for cooking, lighting and transportation.

Another myth that is widespread among energy promoters and policy makers is that:

- 'productive uses are the only way of implementing sustainable energy schemes' – in most cases productive uses are not sustainable unless a range of other aspects are in place, such as access to markets, transport, good management, and so on.

A number of views considered by some to be myths can in fact be viewed as reality. All three of the sources cited above believe that energy can be paid for by the poor. However, all the evidence suggests that there is truth in the statement 'the poor cannot pay'. The World Bank suggests that urban dwellers only switch to modern energy services when they have an income above $3 per capita per day, while the energy poor are on per capita incomes below $2 per day. Also the experience of energy access in Europe and the USA shows that these societies undertook lengthy subsidy and promotion programmes despite having much higher average income levels when energy access programmes arrived. This explains why the switch to modern energy services does not happen unsupported, it requires government investment, political willingness and other conditions. Unfortunately this argument leads people to think that the poor can afford to pay (including even the initial investment) and lends towards the belief that markets forces can solve the problem of energy access.

The collection of cooking fuels in rural areas as free of cost can also be considered a reality rather than a myth. Many people believe that it is simple to redirect the time that rural people would save if they didn't have to collect fuel wood to other more productive activities. This is not realistic, because in most cases extremely poor people simply do not have the opportunity to engage in alternative productive activities either because they do not have the physical or the financial capital. Many people in rural areas live in a subsistence economy and they therefore have no surplus to sell or to process, or if they have any surplus the markets are so far away that they may use other forms of preservation. For example, a small-hold farmer uses his excess cereal production to fatten a pig. The reality, then, is that although time may be saved when people no longer have to collect firewood, this time cannot always be converted to money which will pay for alternative fuels.

Therefore although there are some myths, such as 'energy is not a priority', some so-called myths are in fact undeniable realities. Of these the most important is the 'myth' that people in extreme poverty cannot pay the initial investment for modern energy such as electricity or liquid and gas for cooking, and therefore they will remain energy poor unless the capital investment for energy is subsidized.

Summary points

- History shows that governments played a very important role investing in the process of rural electrification and access to cooking fuels in developed countries. The markets have been welcomed but have not been the driving force.
- The process of rural electrification in the USA took about 30 years.
- When rural electrification was implemented in USA and Europe, the rural inhabitants were much wealthier than the inhabitants of developing countries demanding electricity now.
- The World Bank estimates that the total switch to modern energy occurs at a relatively high income of around $3.3 per day per capita.
- Electricity reforms have followed a particular standard menu. Although most cases may be considered to be technical successes, often they have been social and political failures and have not benefited the poor.
- GATS relies on market mechanisms as the driving force for the provision of services. This is not the best way to reach the poor.
- Overstretched national budgets mean that energy access for the poor is a low priority.
- In rural areas, dispersed populations makes investment costs for grid extension and for small decentralized energy generation high.
- In urban areas modern fuels are available, but poor people cannot afford to pay their full cost.

People in extreme poverty cannot pay the initial investment for modern energy therefore the capital investment should be subsidized.

CHAPTER 3
Energy solutions for the poor

Strategies for overcoming the barriers to energy access

Although the barriers to extending modern energy supplies to rural and urban poor populations in developing countries are considerable, examples exist of projects with promise. Although capital and recurring costs for decentralized energy supply systems can be high, sometimes costs can be kept down when there is national capacity for manufacturing and installing systems (see Table 3.1). Local contribution in the form of labour can also bring down costs. Practical Action has been working since the late 1970s on the promotion of energy access for the poor, mainly in two areas, cooking and access to electricity. With regards to cooking it has been promoting efficient cooking stoves, while in relation to electricity it has been promoting small-scale renewable energies. Annex 1 summarizes Practical Action's experience in this field.

As argued in the previous chapter, one of the barriers to gaining initial access to energy services is poverty, but, even when the initial investment has been covered, what often prevents services continuing to operate in poor countries is lack of affordable, manageable technologies. This includes a lack of capacity in communities to design and implement energy systems, and more importantly a lack of users' capacity to manage, operate and maintain energy systems. Lack of 'energy literacy' also plays a big role in many cases, as there is not sufficient knowledge to understand the benefits, costs, sustainability and risks of energy. To tackle all these issues there must be appropriate policies and strategies.

These strategies are incorporated into Practical Action's approach to energy access programmes which involves more than simply installing technologies. A number of activities are necessary: the assessment of needs, resources, technology options and their viability; technical research and technology development; social research; technology transfer; pilot systems; post-implementation evolution; the design and implementation of credit schemes; and policy research and advocacy.

The long-term viability of energy services for the poor has been one of the most important issues during the last three decades, especially regarding small decentralized energy systems for rural electrification and access to efficient biomass cooking stoves. Practical Action used a straightforward strategy for identifying barriers, and proposing and implementing solutions (see Table 3.1).

For example, with regards to the development and transfer of technology, Practical Action played a decisive role in building capacity for designing, manufacturing and implementing micro- and mini-hydro schemes so that organizations and communities in Nepal, Peru and Sri Lanka have the in-house capacity to design and install these systems. They also manufacture most of the electromechanical equipment and components and even export some components to other countries. Practical Action was also instrumental in the development of 'national standards' for the manufacturing and implementation of these technologies in Nepal and Sri Lanka. Practical Action played a similar role in the development and dissemination of the ANAGI stove in Sri Lanka; this stove is sold commercially in this country and elsewhere in South-east Asia.

With regards to credit schemes, Practical Action has been actively advocating for access of the poor to financial resources, and has developed and implemented a number of credits schemes. Among them is the credit scheme for LPG cookers and gas bottles implemented in partnership with the Women's Association in Kassala in Sudan. This scheme is now being replicated in Al Fasher in Darfur, and by the Micro Hydro Revolving Fund in Peru. The latter is aimed at small-hold farmers, organized rural groups and local authorities, and it provides soft loans with technical assistance for the installation of micro-hydro schemes.

Regarding local capacity, Practical Action has developed tools and methodologies that enable people to participate throughout the process of design and implementation of the schemes, and also creates capacity within the community to ensure future operation and maintenance. A number of organizational and management models have been developed and tested for different technologies and countries. In Sri Lanka, for example, Practical Action has successfully developed and disseminated the 'energy consumers model', while in Peru one of the most successful organizational models for the operation and management of energy schemes is the 'small energy enterprise model.' (Sanchez, 2006a)

Table 3.1 Solutions to achieving sustainability of decentralized energy systems for the poor

Barriers	Solutions
High investment costs of energy technologies (non-affordability)	• development and transfer of appropriate technology; • building the national capacity to design, manufacture and install energy systems and use energy rationally; • development and promotion of appropriate standards; • importing mass-produced equipment (e.g from China) might help in some cases.
Lack of access to financial resources	• development and implementation of credit schemes, with technical assistance and with public and private participation.
Lack of local capacity to operate and maintain energy schemes	• building local capacity to operate systems; • develop and disseminate appropriate organizational models.
Lack of appropriate legal framework	• advocacy for policy and strategy changes at national and international levels; • promotion of energy literacy among the users.

Concerning appropriate legal frameworks, Practical Action advocates for pro-poor national, regional or local regulations that can either, directly or indirectly, contribute to opportunities for greater local participation in the ownership and management of energy schemes and/or enhance local investment. In Kenya, for example, the implementation and demonstration of pico-hydro schemes in isolated communities have served as platforms for discussion and a change in regulations of standalone schemes. In Peru, Practical Action is contributing to changes in national regulations related to renewable energies and rural electrification laws.

Appropriate energy options for the poor

A full energy mix should be considered for poor communities with different emphasis according to location and opportunities. For example, grid and non-grid solutions for electricity supply may be combined with liquid and gas fuels, as well as biomass for cooking where appropriate. For urban inhabitants, grid extension for electricity supply and liquid and gas fuels for cooking are often the most appropriate solutions, although in many cases, biomass for cooking in smaller towns may still be required. For rural inhabitants, the most appropriate option may be a combination of grid for those living close to the transmission lines, with decentralized renewable energy options (including solar PV, micro-hydro and small wind) for off-grid electricity supply. For cooking biomass is likely to remain the main fuel option for the majority, though more efficient and cleaner cooking devices may be used.

Grid electricity has been the most important means of increasing electricity access in the past, including in urban and peri-urban areas. Grid electricity is a fully developed, proven option and its use is regulated and standardized worldwide. However, for rural areas and especially for isolated areas this option becomes economically uncompetitive compared to standalone options. For this reason this book does not deal further with grid electricity, as it is a well known option that will continue to be important where it is economically competitive.

There have been a number of efforts to use standalone electricity generators in developing countries. According to REN21 (2007) by 2007 more than 2.5 million households in developing countries were receiving electricity from solar photovoltaic (PV) systems. The most growth of this technology has occurred in Asian countries (China, India, Bangladesh, Sri Lanka and Thailand), the largest, China, installing 400,000 household systems in the last decade. The technology is also growing fast in Thailand, which has provided electricity with off-grid systems to 200,000 households between 2003–6.

Solar PV is not the only solution for off-grid energy access and it can be costly and provide a limited power supply. Too great a focus on solar PV (driven by international, multilateral corporations) has meant that other viable technologies – wind, micro-hydro and biomass, which can be locally developed and manufactured – are hugely neglected and seem to 'fail' because there are limited financial resources available for large dissemination programmes.

The next section describes a number of energy options for both cooking and electricity supply, some of them more appropriate and used in rural areas and others more in urban areas.

Decentralized electricity generation options

There are two electricity generation and supply options: grid and small decentralized systems (off-grid). In the past, rural people in many developing countries have gained access to electricity by grid extension. However, as the coverage progresses from urban to peri-urban areas the population becomes more scattered and the grid has to reach wider to compensate. For this reason, extending the grid is becoming too costly to provide electricity to the remaining villages and communities because of the higher investment per family required, and the high operation and maintenance costs. This is particularly true in developing countries with higher electricity coefficients (the percentage of rural people with electricity), such as Mexico and Chile, or in low populated countries or regions within a country, such as the Amazon regions in Peru, Bolivia and Colombia where the population is about two or three people per square kilometre.

In this situation, the unsupplied urban population, which by 2005 accounted for nearly 300 million worldwide (see Table 1.1), as well as the people moving from rural areas as a consequence of urbanization, would be best served by electricity from the grid. However, for the majority of rural people living in isolated communities and villages, small decentralized energy systems have proven to be the most appropriate technical and financial solution for electricity supply.

The more successful rural electrification schemes such as in Costa Rica, Brazil, Thailand and Tunisia (Barnes, 2007) have a dedicated rural electrification agency, have used cost reducing electricity distribution technologies, and have engaged communities in the planning and delivery of the electricity services. In addition, these initiatives have developed electrification alongside rural development strategies, thereby building energy end uses for education, health and economic development, and thus improving the viability of this provision. Many developing countries, however, have failed to reach the majority of rural people with grid electricity, with countries in sub-Saharan Africa supplying less than 10 per cent of their rural population with electricity. Even where grid extension has been more successful, there is general agreement that decentralized energy technologies are most appropriate for reaching isolated communities.

Among the most common decentralized energy options are: those based on renewable sources of energy, such as solar PV systems; small hydropower plants; small wind electricity generators (SWGs); biomass (biogas digesters and gasifiers); and small diesel generator sets.

One of the most important advantages of small decentralized systems is that they can be sized according to the specific needs of a community and they allow the involvement of the users at all stages of implementation, encouraging control and ownership by the users. However, these systems generally require large upfront investments, for example, in isolated rural areas the investment cost for a 50 W_p solar PV system is over $1,000 and generates about 5 kWh per month, and a 100 W wind machine costs around $800–$1,000 and can generate about 20–30 kWh per month. These systems also require a number of conditions for them to work, such as local technical assistance and the existence of a market in spare parts.

Small hydropower schemes

Small standalone hydropower schemes are generally run off river systems. Water is captured through water diversion and conducted through a small channel for a distance to create a drop to produce the necessary power. At the end of the channel a specially designed tank (forebay tank) connects the channel to the penstock, which takes water into the power house and brings it into the turbine housing and rotates the turbine, the turbine turns the electrical generator which finally produces electricity output.

During the last two decades there has been intensive research and development work on small hydro schemes, contributing greatly to cost reductions on manufacturing, installation, and to widening the available design options and power ranges. At present the market offers small units from hundreds of watts to hundreds of kilowatts at very low cost ($2,000–$4,500 per kilowatt installed, this cost includes, civil works, equipment, distribution wiring, and transmission lines when required). This makes hydropower very competitive

---

Box 3.1 Micro hydropower in Nepal, Peru, Kenya and Sri Lanka

Practical Action has played a significant role in establishing community-based micro-hydropower plants (off-grid schemes under 100 kW) with communities in Nepal, Peru, Sri Lanka, and Kenya (in this last case these are demonstration projects). These systems, which are designed to operate for a minimum of 20 years are usually 'run-off-the-river' hydro systems. They do not require a dam or storage facility to be constructed, but simply divert water from the stream or river, channel it into a valley and 'drop' it in to a turbine via a penstock (pipeline). Experience shows that with community capital (in labour and cash), financial credit and good management models, these schemes can be economically viable and sustainable. Besides providing power for domestic lighting and basic community services (energy for health centres, schools, communication facilities), village hydro schemes can also be used for running small businesses and income generating activities like grain milling, agricultural tool repair and welding depending on the needs of the community.

Practical Action has found that success factors include: building local technical capacity to design, plan and install hydro schemes; local manufacture of equipment and parts; delivery of maintenance services; community involvement in planning and implementation of the schemes; appropriate financing models to allow access for lower income communities; and proper organization and management.

Photo: Luis Rodriguez

This 35 kW micro hydropower plant was installed in 1989 by Practical Action and the Cooperative Atahualpa, in the North Andes of Peru.

with other energy options. The development of pico-hydro during the last 20 years has made available hydro schemes with the technology to generate hundreds of watts, or even less, at competitive costs. This has been thanks to the use of non-conventional technologies, such as the use of motors as generators with specially designed electronic load controllers. For example, in Asia and Latin America the cost of the equipment (turbine, generator and controller) for a 5 kW pico-hydro scheme costs between $3,000–$5,000, the same equipment in Europe, USA or Japan can cost five times this amount.

Currently small hydropower technologies are one of the most mature small-scale technologies for energy generation in developing countries, but there are also a large number of countries which lack manufacturing and implementation capacity. Hydro-energy schemes are one of the most appropriate technology options for mini-grid electrification in small villages and communities where water resources exist and the topography is suitable.

Small wind energy

Wind energy has been used since early times for grain milling, and during the pre-industrial revolution for pumping water for irrigation and drainage. Nowadays wind energy is one of the most promising energy technologies, and wind generator technology has become popular for electricity generation. It is one of the fastest growing modern electricity generation technologies and contributes to the energy security of developed countries, through the implementation of large wind farms which feed the grid. Although it has good prospects for the future, large wind energy systems and wind farms are still not contributing significantly towards overall energy needs in developing countries. The exception is China and India which are fourth and fifth in the world in terms of the amount of installed wind energy capacity, though in both cases most of that capacity feeds the national grid.

In developing countries, in the past, small wind power technology has been promoted much more for wind pumping rather than for electricity generation. As a consequence, there are a few countries with some local technological capacity to manufacture and install wind pumps, but only a handful of developing countries with the capacity to manufacture and install small wind turbines, among them, China, Brazil, and more recently the Philippines, Peru and Sri Lanka.

At present small wind turbines below 5 kW are becoming more attractive for small application in rural electrification, especially for dispersed families, for small facilities and/or services, such as health, education, communications and others. However, there is still the need for technology transfer on manufacturing, design and implementation of small wind turbines in developing countries in order to reduce costs. Practical Action's experience in the development and promotion of this technology is that small systems below 500 W can be manufactured at a third or less of similar products from developed countries, but still with high reliability and good performance. A 100 W wind

Photo: Jose Chiroke

Trained members of the community install a 100W micro-wind energy system at 'El Alumbre', in the North Andes of Peru, with assistance from Practical Action Peru.

electricity generation system based on Practical Action's technology is being commercially manufactured in Peru and sold at about $600. A similar unit from the USA or Europe costs $2,000–$3,000. Practical Action is currently promoting three sizes of wind turbines for electricity generation – 100 W, 200 W and 500 W – the first two models are appropriate for household electricity provision, while the third can power social service facilities such as schools, health centres, and communications facilities. Slightly larger wind turbines (from 10–100 kW) are appropriate for hybrid systems (e.g. wind-diesel) for small towns and villages.

Solar energy

Solar energy technologies, including solar PV and solar water heating, have shown potential in energy access for the poor. They can also be combined with other technologies for uses such as cooking and drying.

As with wind energy, solar PV has had great success during the past decade with huge annual growth. It is now starting to play a role in energy security in developed countries, while in developing countries it is still in the phase of piloting and promotion. However, the costs are still very high compared with other options (see Table 3.2).

An interesting feature of solar PV is its modularity. Thus, it has the potential for meeting small electricity needs of rural isolated homes or small institutions, where other energy sources do not exist or are unsustainable.

Despite its costs and limitations on energy generation, solar PV systems are being heavily promoted in developing countries by a range of institutions, including large agencies like the World Bank, UNDP, EU and others.

Solar photovoltaic panels are erected in Peru.

Box 3.2 Solar PV for developing countries

The most common off-grid solar PV system is the solar home system (SHS). That is a solar PV array of between 20 and 50 Wp, with a control system, battery storage and connection to domestic appliances, usually high efficiency lighting plus one or two power sockets. (Wp, or Watt peak, is used in solar PV systems to mean the wattage produced when the system is generating at its highest efficiency and under a radiation of 1000 W/m²).

Unless the SHS is heavily subsidized, it is generally not affordable to lower income rural homes. One way of reducing costs is by subsidizing the system and reducing its size to below 10 Wp (AEPC, 2006). In addition, there needs to be services in rural areas for parts and servicing of the solar technology. There are an increasing number of innovative enterprises which aim to overcome the barriers faced by poor communities in accessing solar PV systems. For example:

Solar lantern with radio connection: Practical Action's consulting subsidiary developed a low-cost solar lantern. The lantern can provide up to six hours of high-quality light, or a combination of light and radio output to bring news and information to households.

SELCO-India (Wheldon, 2005) shows that solar PV systems can make solar energy affordable yet commercially viable. SELCO is a private business, based in Bangalore, which provides SHS and other solar services to low-income households and institutions. Its network of local sales and service centres are set up where microfinance organizations can provide loans to customers. All systems are sold on a commercial basis, but SELCO is committed to providing the highest quality services to poor people on financial terms they can afford.

Barefoot College, India (2003). During the last ten years this project has been introducing solar technology to remote and inaccessible villages in the Himalayas. Run by the Barefoot College in Rajasthan, India, the project has shown that with appropriate training, poor and rural communities can install solar equipment in their villages and maintain it without any further external help. The project has trained illiterate and semi-literate villagers as 'Barefoot Solar Engineers' at its Barefoot College in Rajasthan. After the training, the participants return to their home villages to install solar units and provide their communities with a skilled and competent repair and maintenance service.

Solar service delivery company (Wheldon, 2005). Sunlabob Rural Energy Ltd, Lao PDR, provides solar power to electrify rural villages at a price people can afford. The success and sustainability of the scheme lies in a rental service which avoids upfront costs and direct subsidies and a network of trained rural entrepreneurs who can respond quickly to any technical hitches in the more remote areas and so maintain the high quality of the PV systems.

A boy completes his homework by the light of a solar lantern

Photo: Jonathan Adjei, Webjetty, Kenya

Biogas

Although biogas is a technology that has been promoted primarily for lighting and cooking, it has also been suggested for electricity generation.

Biogas is generated from the decomposition of organic materials in a confined environment (the biodigestor) and without the presence of oxygen. It has been used for many decades as a cooking and lighting fuel and in the early part of this decade, in China alone, there were about 7 million units (Nijaguna, 2002). In India, Nepal and other developing countries there have also been important dissemination efforts on small-scale biogas plants in rural areas, mainly for household energy supply (see Box 3.3).

An important issue, and the biggest barrier, for biogas technology is the fact that it requires certain conditions to install and run a biodigester. These conditions do not match the realities of daily life for most poor people. For example, the 10 cubic metre Chinese model requires a minimum mass of raw material equivalent to the dung produced by at least two large cows in order to produce enough gas for lighting and perhaps make breakfast. Another important requirement is the handling of the raw material which sometimes may not be culturally acceptable to the communities.

Box 3.3 Biogas Sector Partnership (BSP) in Nepal

The Biogas Sector Partnership (BSP) in Nepal managed the installation of over 150,000 domestic biogas plants in Nepal from 1992–2005 (Boyd, 2006). The plants use cattle manure to provide biogas for cooking and lighting. In addition, about 75 per cent of the plants incorporate toilets. The programme uses a combination of part subsidies on the cost of installing the biogas system with quality assurance checks on the manufacturers to ensure a high level of success.

About 80 per cent of the 4.2 million households in Nepal use fuel-wood, cattle-dung cakes and agricultural residues for cooking, and kerosene for lighting. Demand for fuel-wood substantially exceeds the rate of regrowth, and this is leading to degradation of the land and damage to vital watersheds. Cooking indoors over open fires and lighting with kerosene gives dangerous exposure to air pollutants and a high risk of fire, particularly for women and young children who spend much of their time indoors. In addition, women and girls have the drudgery of collecting fuel-wood, which typically takes three hours each day.

The biogas plants replace nearly all the use of fuel-wood and make cooking easier, cleaner and safer. In 20 per cent of houses biogas provides safer lighting as well. This saving of unsustainable fuel-wood use also reduces carbon dioxide emissions. The provision of toilets improves sanitation and the effluent from the biogas plant is a valuable organic compost.

Small diesel sets

Over the past decades, small diesel sets have been one of the technologies used most often for rural electrification. Although they are currently much less promoted, they are still used to supply electricity to small towns and villages in rural areas in developing countries.

Diesel sets have become popular in the past because, among other reasons, these systems require smaller initial investment than any other renewable energy system (equivalent in size), and require less time for planning and implementation. Currently diesel sets are becoming less popular in rural electrification programmes, partly because of environmental concerns (CO_2 emissions) and partly because of their high running costs. For example, a study in India showed that electricity generated from diesel sets costs about three times that obtained from the grid (Cust et al., 2007), especially when there are no subsidizes for fuel and maintenance costs. Experience in different parts of the world shows that small diesel sets only run for a few hours a day (Luque and Hegedus, 2003; Sanchez, 2006b; Gül, 2004) and can be stopped for several days when failures occur or when fuel is not available.

Diesel sets, however, may be attractive and appropriate for some specific and well-tried services, for example, multifunctional platforms in some countries in Africa (Brew-Hammond and Crole-Rees, 2004).

Other options for rural electrification

This category includes some older technologies which have yet to be fully tested as small-scale power producers for developing countries, as well as new innovations which show good potential. Technologies currently in the process of development are at different stages; some of them have reached the stage of piloting and have even been installed in some units to provide energy services, others are still in the development phase, and some are proposals for development. The different technologies and their state of development are as follows:

Gasification and pyrolysis. Gasification and pyrolysis are biomass-based technologies which can produce a gas for fuel; both processes are used separately and operate at high temperatures and in the absence of air. The gas may be used in existing boilers or furnaces or can fire combustion turbines to generate electricity (BREDL, 2002).

Fuel cells. This technology converts fuel to electricity via an electrochemical process; unlike a battery, the chemical input is not stored in the system, but is fed continuously into the fuel cell. The chemicals in the fuel cell are hydrogen and oxygen. Various fuels, including natural gas, methanol, ethanol and gasoline can be used (reformed) to provide the hydrogen necessary for the fuel cell.

The technology of fuels cells has not yet been promoted for rural energy applications. It appears that there are no technical barriers to this use, however there is no evidence yet about its suitability in terms of costs or social and cultural acceptability.

Thermoacoustic systems. The generation of electricity using the principles of thermoacoustics is a relatively new concept. Recently researchers from a group of UK universities (Nottingham University, Imperial College London, Queen Mary University London and Manchester University) and Practical Action collaborated under the SCORE project to develop a new type of machine, which could deliver electricity, heating and cooking. The project is in its early stages. The attractiveness of this machine (assuming successful results are achieved) is that it should be possible to make very small units, tens or hundreds of watts, which could be used for household energy purposes and could provide at least two of the very basic energy requirements, lighting (with electricity) and cooking. The cooking system promises to be more efficient than devices such as the three-stone open fire used currently by the poor.

Modern biofuels. As in the case of the biogas, biofuels have primarily been used by the poor for cooking. However, in recent years there has been growing interest in biofuels to replace high carbon fossil fuels. Biofuels include the whole range of fuels (solid, liquid or gas) extracted from organic products. When people speak about modern biofuels they are referring to oils from seeds and plants, or to alcohols (extracted by fermentation) from a variety of agricultural products. At present biofuel is one of the most promoted energy technologies. Large organizations such as UNDP and the EU have already established promotion programmes. The EU has set targets on the use of biofuels as part of their energy mix.

Although there are some efforts to use biofuels in developing countries for household energy applications such as cooking, electricity and mechanical power, there are no clear indicators about the possible contribution from modern biofuels to energy access by the poor. From the available information on production costs, agricultural practices and market issues, it looks as if biofuels will contribute primarily to energy security in developed countries rather than to energy access for the poor. The organizations promoting biofuels believe that this energy source will contribute greatly to energy access, income generation and employment for the poor, however, all these expectations are yet to be proven.

So far, promotional activities have concentrated on producing large quantities of biofuels for export. In Africa the planting of *jatropha* is being widely promoted, palm oil is popular in other countries, especially in Asia and at a smaller scale in Latin America. However, there are few or no activities aimed at energy access for the poor based on small-scale biofuels. GVEP are currently supporting projects in local ethanol production from agricultural wastes, e.g. coffee fruit, but these are very much at the early stages.

Energy options for cooking

For cooking three main options can be considered to improve the energy access of the poor:

- improve the efficiency of cooking devices using solid fuels (improved cooking stoves);
- produce and use a range of liquid and gas biofuels for cooking;
- switch to liquid and gas fuels (use of fossil fuels).

Efficient cooking stoves

Over the last three decades a large number of models of efficient cooking stoves have been developed and tested worldwide. However many of these models have only reached the pilot and demonstration phase. Only a few have been produced commercially. Many dissemination programmes have also failed. The most important reasons for failure appear to be:

- failure to deliver cooking technologies which match social and/or cultural cooking practices;
- programmes have been short term, donor driven, and in some cases stoves have been given away to potential users. When such programmes end they do not leave the capacity to manufacture, sell or maintain stoves;
- in some cases the stove models have failed technically, for example when users had difficulty in operating the stove or when the stoves saved them little fuel.

A review paper from the World Bank (Barnes et al., 1994) revealed the common features of successful programmes involving stove dissemination: the programmes focused on a group of users who were most likely to benefit from and consequently adopt the improved stove; the stove itself is minimally subsidized; there is significant interaction between the people who design, produce and use the stoves; the stoves are mass produced; and external support is limited to the development, production and distribution of stoves.

However, there are some very notable success stories which provide useful lessons for new stoves programmes. The most successful stoves programme has been in China, where more than 180,000 improved stoves had been introduced by 2000 (Edwards et al., 2004). The success in China has been attributed to stove designs suited to users' needs, targeted national promotion schemes and effective local implementation, including setting up commercial rural energy companies.

Although many of the early stoves projects achieved little replication, there have been a few successful programmes. The ANAGI stove in Sri Lanka and the Upesi and Jiko stoves in Kenya are success cases in which Practical Action has been involved. In these cases efficient cooking stoves were developed and reached large numbers. These stoves programmes were developed with commercialization in mind from an early stage and there has been an important focus on the issues of market, replication, mass production, low cost and efficiency, among others. Each model was developed through a pilot project, including design and development, technology transfer, and local

---

Box 3.4 The Upesi stove, Kenya

As a product the Upesi stove has had a widespread impact in rural Kenya (Practical Action, 2004). The technology is relatively straightforward. What makes the Upesi a success is the way in which non-technical issues, including the use of the market, the participation of the community, and entrepreneurial training and skills development have been addressed as part of the development approach to the technology. Practical Action worked with three groups of women potters to design and test prototypes.

The Upesi stove was selected for production after field trials that showed it could provide fuel wood savings of up to 43 per cent compared to a three-stone fire, and has a life span of around four years. Some stoves have reportedly been used for up to ten years. The community is actively involved in the manufacture and promotion of stoves, which are sold commercially on the open market. The Rural Stoves West Kenya project has trained 13 women's groups (approximately 200 people) to make improved stoves. As well as production training, the women gain skills in business management, including marketing. The training is participative. Women identify their own training needs, devise the programmes and control their pace. The annual production is estimated at 10,000 to 11,000 stoves, which are sold mainly in the peri-urban areas and the profit generated by the stoves provides artisans with employment. As a result, the women involved have gained status, self-confidence and financial independence.

The technical training of the women, by itself, would not have been sufficient to achieve success. Training in group organization, management, marketing and business skills was crucial, particularly because the aims of the project focused on the benefits to producers and the development of a commercial market for stoves.

Production of ceramic Jiko stoves by the Keyo Women's Group established in 1984, Kisumu, Kenya

Production of the Anagi stove in Sri Lanka

Box 3.5 Anagi stove, Sri Lanka

The Anagi model (INFORSE-Asia, 2006) was designed to be disseminated through commercial channels. It was initiated through a joint project of Practical Action and the Ceylon Electricity Board, during the mid-1980s. The success of this model was immediate and from 1987–90 there were around 80,000 units sold. This initial success encouraged Practical Action in 1990 to embark on a joint project with a small, new local NGO called IDEA (Integrated Development Association), with the purpose of making the Anagi cooking stove widely available throughout Sri Lanka. This project enabled a number of potters and artisans to become producers of this stove through training activities. The project also included activities on marketing, which were partly through training to manufacturers and partly through conventional media information dissemination.

By 1992 about 110,000 units were sold in Sri Lanka (Amerasekera, 2006). The ARECOP (Asia Regional Cook Stove Program) in its online presentation of publications referring to the Anagi stove manual states, 'Anagi is a two pot-hole pottery improved stove, originally developed and disseminated in Sri Lanka. Anagi has been one of the most successful pottery improved cooking stoves in Asia, with more than half a million units being disseminated (mostly through commercialization) since 1991. 'A number of factors led to the success of the Anagi stove (Amarasekara and Atukorala, 2003):

- There were multiple reasons for introducing the improved stove: national concerns over deforestation and lack of fuel wood; aims to reduce fuel wood consumption in the house, and reduce the burden on women; to improve household conditions though cleaner kitchens; and to increase incomes for the potters' groups who manufacture the stoves.
- The programme had government support at a national level for stove development, with local NGO implementation.
- The programme has three distinct phases: stove design and piloting; small-scale dissemination; large-scale dissemination and full commercialization.
- Technical development and considerable local field testing meant that the Anagi suited both local manufacture by potters and also suited the local cooking practices.
- Improved quality assurance on the product gave greater consumer confidence.
- National media promotion plus local promotion by the producers and retailers led to rapid wide-scale dissemination of the stove.
- A key objective was to move from a subsidized programme to a fully commercialized product. This was fully achieved in Sri Lanka.

---

production and commercialization. Each project arrived at a simple, small, efficient model, which can be mass produced locally and commercialized.

Switching to liquid and gas fuels

Switching to kerosene and gas for cooking is an important option for urban families. In urban areas there is also a wider choice of cooking fuels, kerosene, LPG, natural gas, wood and charcoal. In many cases the access of the poor is limited, either due to the cost of the fuel or due to the cost of the cooking device. The case from Sudan (see Box 3.6) demonstrates that such barriers can be overcome.

For rural areas the option to switch to liquid or gas fuels is difficult due to the limited transport existing in most developing countries.

Box 3.6 LPG cooking in Sudan[1]

High dependence on biomass fuels for household energy not only contributes to environmental degradation, but equally causes serious health problems, especially for women and young children. Practical Action Sudan initiated the first work of its kind in Sudan by monitoring household indoor air pollution (IAP) levels, using participatory research methods with a sample of 30 poor households from a semi-urban residential area, Wau Nour, Kassala. Traditionally, cooking takes place on inefficient three-stone stoves using firewood. The IAP monitoring revealed high levels of particulate matter and carbon monoxide.

A participatory discussion with household members revealed a high awareness about the health risks of smoke and about the possible interventions to reduce or eliminate smoke. Almost all households claimed that LPG is the best way to mitigate the smoke health risk. However, the issue of accessibility to LPG cylinders and burners was raised as the dominant constraint (Hood, 2006): the cost of buying and installing these ranges between SD9,200 (about $45) and SD17,900 (about $85).

Practical Action has helped the Wau Nour sample households switch to LPG by organizing loans to cover the cost of changing to LPG. Using a revolving fund, the project enabled 167 households to switch to cooking with LPG, which reduced levels of IAP in the home by more than 80 per cent. A scaling-up strategy was developed with the Women's Development Associations (WDAs) and other partners and stakeholders. This enabled women to buy ovens and gas cylinders by establishing a revolving fund, which was managed by the women who also contributed to it financially. The gas agent in Kassala agreed to supply cylinders on an instalment basis. Based on energy savings and conservation calculations, households could repay the cost of LPG appliances in a maximum of six months.

Now over 925 households have switched from using wood to gas and have acquired all the necessary appliances. Refilling outlets were established in Wau Nour and Kadugli residential areas to give easy access to gas. A new plate for baking Kisra (a thin sorghum pancake which is the main food in central Sudan) is being manufactured locally. Training in safety measures is undertaken through the involvement of the Civil Defence Forces, and so far no accidents have been reported.

Photo: Practical Action

LPG cookers have been introduced in El Fasher, Darfur, Sudan, through micro credits

Advantages, disadvantages and costs of different energy options

Electricity options

During the last two decades there has been enormous technological progress on most renewable energy generation systems: solar PV, wind electricity generation systems, and more recently biofuels. There have been important incentives on their use through promotional policies and subsidies, which have stretched the commercial sector and encouraged huge growth rates in these technologies: world wide, solar PV and wind technologies are growing at around 25 per cent yearly and biofuels are getting close to these rates. These technological and commercial pushes have also contributed to sizable cost reductions, for example, solar PV production costs ranging from $0.28–0.90 kWh (Mutschler, 2007; Turkenberg, 2000; Ruoss and Taiana, 2000) and wind generation costs from $0.06–0.18 kWh (European Commission, 2005). Biofuels have also experienced similar low costs. The main drivers for progress in the commercialization of these technologies have been the need for energy security in the north and the need to mitigate the global threat of climate change due to greenhouse gas emissions.

However, this progress in cost reduction has contributed little to energy access for the poor, and the costs of small standalone energy generation systems (required for rural areas) still remain high. For example, studies on the cost per kilowatt hour from solar PV for mini-grids in Sudan is reported to range from $0.96 to 1.62 kWh (Croxford and Rizig, 2007) , yet family home systems are much more expensive and can be easily over $3.0 kWh depending on different factors (e.g. radiation, cost of equipment, installation, operation and maintenance, among others). In Argentina isolated households who receive electricity under the rural electrification concessions programme, are charged from $8–$10 for about 3 kWh per month despite government subsidies of 50 per cent of the cost (Reiche et al., 2000). In the Dominican Republic a rural energy programme that rents out 50 W_p systems charges a monthly 'fee-for-service' of $15–$20 monthly (Martinot et al. , 2000). For very small wind systems energy costs are high, generally above $1 kWh.

The experience of Practical Action is that building local capacity to design, implement and run small generating systems can cost much less than what is often reported internationally, especially for those technologies which can be manufactured in-country (e.g. small hydro or small wind generations). Furthermore, if the implementation and post-installation technical assistance is provided by local technical assistance the unit cost of energy comes to a fraction of the cost reported internationally.

Table 3.2 shows different energy options for isolated rural areas, the cost and availability. The costs for technologies based on local resources – solar, small wind, small hydro, and biomass – are calculated under the assumption these technologies are manufactured and installed using local capacity, while the equipment for solar PV systems and diesel sets are imported from developed countries but installed using local capacity. For the purpose of calculation of energy costs, the life span estimated for the different systems were hydro, wind and solar PV systems 20 years and diesel sets 8 years.

The important issue to point out is that despite the high generation costs of energy from small decentralized systems, these systems are still competitive with the grid and acceptable to the poor. In most cases these are the best options or perhaps the only option if the people want access to better quality energy compared with existing sources.

Table 3.2 Some features of availability and cost for energy sources for rural isolated areas

Source	Availability	Cost[2]	Technical issues
a) Fuels for cooking			
Kerosene	scarce, unreliable supply, problems of transport	high cost of fuel and cooking devices	cooker and spare parts may be found in the cities
LPG	hardly available	very high cost of fuel and cookers	cooker and spare parts only in large cities
Solid biomass (wood, dung, agricultural residues)	generally available	low (but could be high if collection time is counted)	low efficiency, smoke emissions
Charcoal	available in some markets, especially in urban areas	high	low conversion efficiency if counted from extraction to useful heat
Biofuels	hardly available for cooking	–	little experience on the use of biofuels for cooking
b) Sources of electricity			
Grid	not available for isolated villages and communities	very high for rural isolated people, but for those living next to grid paths it may be competitive	grid connection is a good solution, when the cost and availability are right
Solar PV	available all over the world	very high ranging from $1.5–3.5 per kWh	limited use due to the high unit cost of energy, inappropriate for productive uses
Small wind generation systems (below 1 kW)	frequently available but not everywhere	high $0.30–$0.80 per kWh	sometimes there may be several days without wind, therefore back-up needed
Small hydro systems	site specific	medium $0.20–$0.60 per kWh	it is the most mature technology, proven technically and socially in the field
Small diesel sets in remote areas	difficult access	from $0.30–$1.20, depending on the intensity of consumption	difficulties operating and maintaining expensive spare parts, noisy and polluting
Biofuels	new source of fuel, still to be proven	expected to be high	long production chain, from farming, processing to energy generation and use (possible social complications)
Fuel cells	not yet available	–	to be proven, but cost of transport and use could prove to be difficult for the poor in isolated communities

Cooking fuels

Regarding fuels for cooking, the cost of the different fuels varies greatly depending on location. LPG and kerosene may not be available in isolated areas, or if they are the cost may be several times the cost in towns, for example, in the remote villages of the Peruvian Amazon kerosene can be three to five times the cost it is in cities.

Delivery of energy for the poor

Despite the fact that there are proven technical options for energy access for the poor, energy delivery to the poor has proven to be very complicated, not only because of people's lack of purchasing power, but also because of the problems of sustainability; energy access promoters are therefore engaging actively in the innovation of delivery models in other areas besides the technical ones, the most important ones being financial and management of small-scale technologies, and implementation of large projects or programmes with the full commitment of governments.

Financial and management models for electricity access

New approaches to delivering decentralized energy to low-income communities are emerging, with mixed results so far. Recent financial models and promotion strategies for energy access have been embraced mainly by financial organizations, such as the World Bank and Regional Development Banks (Asian Development Bank, African Development Bank, Inter American Development Bank). The investment for these activities has been mostly met by bilateral aid organizations.

A World Bank article (Reiche et al., 2000) argues that off-grid technologies are sufficiently developed and piloted, therefore sustainable markets are now required which persist beyond the development assistance phase. They describe four types of emerging service delivery mechanisms being piloted by the World Bank. There are several examples of projects which can be grouped under these four categories, some of them have been designed and implemented with direct intervention by the World Bank and its partners, but there are also several examples designed and implemented by other agencies and the private sector. The four models are described here along with examples of where they have been implemented.

Decentralized virtual utilities

Operating on the same principle of 'fee-for-service' as traditional utilities, these examples rely on the implementation of dispersed technologies (such as solar home systems) installed on the premises of the users who are charged a monthly fixed payment. It is believed that the same principle can be applied for other arrangements and technologies, for example, micro-hydro, wind energy or even diesel sets.

Among the best known organizations working with this model of energy service delivery is the company SOLUZ, which began activities in the Dominican Republic and afterwards expanded into Honduras. In the Dominican Republic SOLUZ has installed solar home systems on the basis of fee-for-service, the systems remain the property of SOLUZ and the company charges for energy use on a monthly basis. In Honduras, SOLUZ reports that it is working with both modalities – fee-for-service, as well as selling equipment. In Honduras SOLUZ reports that it has installed 1,600 solar PV systems, out of which 1,100 are under the fee-for-service model, and they recognize that this model is becoming more popular in Honduras. For the fee-for-service model, the systems range from 20–100 W_p at monthly costs of $10–$20, while for the other modality (the sale of the system) they provide systems ranging from 30–100 W_p at costs of $600–$1000 per system.

By 2003, in the eastern province of Zambia, three solar energy service companies have been operating for more than two years, with 400 solar PV installations. The companies remain the owners of the equipment and charge a monthly fee-for-service. The fees cover the full operational costs incurred by the companies, including battery replacement, making them independent of further support from government or donors. The installed equipment is a standard size of 50 W_p, with four lights and a socket for radio, light or other DC appliance.

In South Africa, NuRa, a small private company managed by local staff implemented a rural energy programme. By 2004 this company had 60 staff members and had installed 6,000 solar PV units on a fee-for-service basis. In August 2004 they signed a contract with the government to get subsidies for the installation of 8,000 solar PV units over a period of two years, by which time the NuRa objective was to reach break even point with 14,000 SHS. However, by January 2007 NuRa lost the contract with the government and the service is no longer viable, as the systems were subsidized by 80 per cent of the total investment. (See Box 3.10 for further information on South Africa's rural electrification programme).

In Morocco, SunLight Power Maroc, S.A. (SPM) was founded in 1998. This company has installed more than 2,000 SHS and its business is based on a permanent presence in rural areas. This is achieved through wholly owned branch offices in the middle-sized towns of Taza, Sefrou and Taounate, in the north-eastern part of Morocco, as well as a regular presence at rural markets. On this basis, SPM offers continuous technical support and follow-up to its customers. They sell, install, operate and help finance SHS. Its core business is the sale of off-grid solar based electricity services on a fee-for-service scheme, which is chosen by approximately 80 per cent of its clients. The monthly payment ranges from $9–$24 (Benallou, no date)

Local electricity retailers

Under this model, small cooperatives establish electricity retail businesses. Electricity comes either from an off-grid system or from the grid and they buy in bulk. The common factor is the ability to prepare sound business plans to obtain credit financing or the backing of a strong partner. India is an example of where this model is being used, with independent rural power producers (IRPP) contracting with communities.

Experience with this category of market approach does not go beyond a few small and isolated local initiatives, such as small hydro schemes installed or promoted by NGOs, institutions or research groups. One of the most relevant examples which has been promoted by Practical Action is described in Box 3.7.

Box 3.7 Revolving fund for the installation of micro hydropower schemes in Peru

Practical Action Latin America manages the revolving fund to provide credit to micro hydropower systems. It offers soft credits and technical assistance to municipalities or small isolated villages, to organized groups of peasants or private individuals who want to install their own small micro hydropower scheme to generate electricity and provide either electricity services or other services that require electricity to operate. The credits are offered only to remote off-grid locations, where the grid will not reach in the near future. The criteria to select the beneficiaries of the loan scheme include not only the technical, financial and environmental viability, but also the potential contribution to the development of the poor through enabling access to other services and creating jobs and income.

The basic conditions are: loan size from $10,000–$50,000, according to requirements and financial evaluations; from 2–5 years recovery period; interest rate of 10 per cent (in hard currency, i.e. $); commercial guarantees required; free technical assistance; and the borrower needs to show an initial evidence of their capacity and willingness to pay back the loan. Practical Action provides technical assistance to evaluate the technical viability, potential contribution to development of the poor, all training needs, and organizational support for managing and operating the systems in a sustainable way.

Currently there are more than 40 schemes installed, most of them in the Department of Cajamarca, in the northern Andes region of Peru. More than 30,000 people (6,000 families) have electricity for household use and about four times this figure have benefited indirectly. Indirect benefits include: better education services through the use of electricity for lighting and providing adult literacy, use of modern equipment for teaching (use of computers and DVDs), evening classes for adult literacy, better health services (24 hour a day provision, new programmes of health, use of modern equipment for the preservation of vaccines, use of dental equipment with new and better instruments), access to communications, a range of businesses and services in the community (for example, repairing agricultural tools, repairing radio and TV systems, battery charging, milling cereals, ice making and others). More than 300 new businesses have been started, such as food processing, coffee shops and stores.

Energy retailers

Most of the examples of energy retailers occur only in a few countries. Decentralized Energy Systems (India) Pvt. Ltd., better known as DESI Power, is dedicated to the socioeconomic development of rural India through the provision of electricity and energy services in villages. The current focus of DESI Power is the promotion of decentralized rural power plants using renewable energy sources and setting up agro-based microenterprises through EmPower Partnership Programmes. In addition DESI Power has built decentralized power stations for supplying electricity to industries and technical institutions. In 2005 DESI launched its programme called the 100 Villages Empower Partnership in the Araia District, Bihar – they plan to install 100 gasifiers for electricity generation in 100 villages.

In 1998 China designed the Brightness Rural Electrification Programme aiming at providing electricity to over 23 million people (Long, 2006). In late 2001, China's State Development and Planning Commission (now the National Development and Reform Commission) launched the programme 'Sending Electricity to the Townships' and in 20 months the programme reached about 1,000 townships in nine western provinces. By 2003 the installation was completed and was composed of 20 MW PV, 840 kW wind and 200 MW small hydropower, for a total investment of $240 million to subsidize the investment cost (NREL, 2004). The government was then in charge of drafting guidelines for tariffs and ownership of the systems. The next phase was expected to be the provision of electricity to 20,000 townships. The programme was accompanied by a number of other activities such as capacity building and efficient and productive use of energy. The programme was implemented through a competitive bidding scheme; the bidding price was different for each zone (taking into consideration the costs due to isolation and degree of difficulty). The next phase is expected to be finished by 2010. For further information see Box 3.9.

Mongolia has one of the lowest population densities in the world and an important proportion of this is nomadic. The government is currently implementing a village electrification programme to install isolated energy systems based on decentralized energy technologies, such as solar PV and small wind systems. The administration of the systems is expected to be done by qualified RESCOs (Rural Energy Service Companies). The programme will be implemented in three years and will reach 100 or more villages.

Energy equipment retailers

Small-scale technologies (solar, wind and others) are distributed through small local dealer networks to penetrate isolated rural areas. The important issue in this model is to create the financing infrastructure that enables the dealers to extend credit to low-income families.

ADESOL of the Dominican Republic provides solar rural electrification through microcredit. This organization was founded in 1992 and started offering solar energy in 1994 thanks to the Small Grants Programme (see chapter five). Grant funding paid for a small number of homes to obtain solar PV panels, these people along with the rest of the village then formed a revolving fund to help others to get panels. After paying a deposit of approximately $115, families could repay the loan at around $6 per month. The revolving fund financed panels for more than 600 families, in 18 out of the 30 provinces of the Dominican Republic. At the same time ADESOL trained rural entrepreneurs and promoted a network of 16 microenterprises for disseminating solar PV. By 2000 ADESOL claimed that as well as the initial 600 units promoted, the 16 microenterprises have been able to sell about 5,000 solar PV systems to the same number of families (Martinot, 2000).

ADESOL argues that this model has been successfully proven in the Dominican Republic, they report low loan arrears of 3 per cent and claim that the model could easily be scaled-up. ADESOL believes that this model is appropriate for people because it helps them to pay the full cost of the system and it will ensure the development of a sustainable market for the technology. They believe in the concept of ownership and responsibility with regards to the use of electricity.

Grameen Shakti (GS) Bank in Bangladesh has introduced a micro-utility model in order to reach poorer people with SHS. Another successful GS venture is the Polli Phone which allows people in off-grid areas telecommunication facilities through SHS-powered mobile phones.

Grameen Shakti has developed an effective strategy for reaching people in remote and rural areas with solar PV technology. It involves: soft credit through instalments which makes SHS affordable; intense grass-roots promotion through demonstrations, fairs, meetings at the local level; community involvement and social acceptance; effective after-sales service; understanding market demand through quality products at minimum costs, product diversification and innovations to maximize benefits. By 2006 the Grameen Shakti reported that they had sold and installed over 65,000 SHS in rural Bangladesh (Ahammed and Taufiq, 2008) and brought major benefits to its users. Grameen Shakti was established by the Grameen Bank in 1996 to promote, develop and supply renewable energy technologies to rural households in Bangladesh. It seeks to improve the livelihoods of people who cannot access grid electricity.

SELCO-India is a private business which has designed and sold over 48,000 SHS. SELCO does not provide credit or loans, but has built up working relationships with local banks and microcredit organizations over many years. This has given finance organizations the confidence to provide credit for solar PV systems, as well as an understanding of the payment terms which different owners may need. Some users work directly with the finance organizations, others work through self-help groups which gives additional security on loan repayment.

SELF manages a small but highly visible solar PV pilot project in the Zulu community of Maphaphethe in the Valley of a Thousand Hills in KwaZulu Natal, South Africa. With a grant from the US Department of Energy and a contract with the South African Ministry of Energy in Pretoria, SELF brought the first SHS to this rural Zulu community in early 1996. The community, of around 12,000 people and headed by a young progressive *nkosi* (chief), provides an attractive future market for a sizable programme of domestic solar electrification in a region that South African utilities do not expect to reach for 5–10 years (if ever). Solar home systems specified by SELF cost approximately $600 and include a 53 W_p solar module, charge regulator, four compact-fluorescent lighting fixtures, wiring, 98 amp-hour deep-cycle battery, switches and mounting hardware. Loan terms are set at current development bank loan rates, with a 10 per cent down payment and three years to pay the balance. So far there have been no defaults and the user pays the full cost of the system without subsidy.

Creative concessions

This model offers the provision of electricity to rural families in an area or region as a concession through competitive bidding. The winner signs a concession contract of electricity services through the lowest cost option and the energy provision is supervised by the regulator and subject to quality and competitiveness requirements.

The only known case of this model is the rural electrification concession in the Province of Jujuy, Argentina, to PERMER which is part of the programme PAEPRA (Programa de Abastecimiento Eléctrico a la Población Rural de Argentina). This programme aims to provide electricity to 1.4 million people living in 314,000 households and to 6,000 public services distributed through 16 provinces, all in areas distant from the power grid (Reiche et al., 2000).

PERMER (Proyecto de Energía Renovable en el Mercado Eléctrico Rural), started in 1995, is the first rural electrification concession project world wide. This project aims to provide electricity for lighting and social communication (radio and TV) to about 70,000 rural households and 1,100 provincial public service institutions through eight private concessionaires, using mainly renewable energy systems. The estimated total costs of PERMER over an implementation period of six years, amount to $120.5 million, funded by a combination of a bank loan (25 per cent), donation from Global Environment Facility (GEF) (8.3 per cent), government subsidies (21.3 per cent), the concessionaires (36.5 per cent) and the customers (8.3 per cent), It is the world's first and most advanced rural off-grid concessionaire, by 1999 this concessionaire supplied 556 rural households and 43 additional schools with individual SHS of different sizes. PERMER now serves a total of 3,050 rural customers, 1,333 of these with individual or collective solar PV systems.

Concerns about the market delivery mechanisms can be summarized as:

- Experience of the market approach has been mainly with solar PV technology; hydro and wind technologies are hardly considered or experimented with, while others such as biomass are ignored.
- Despite the apparent long duration of the experiments (since the 1990s or earlier), there is insufficient confidence in any of the approaches. The Grameen Shakti Bank appears to be one of the most relevant cases;

however it still looks to be site specific for densely populated regions of Bangladesh.

• With regards to finance all cases except for the Grameen Bank have required significant subsidies to work.

• The cases that have shown fast results are led by government, for example, in China, Brazil (see Box 3.8) and Mongolia, where the government is subsidizing the total initial investment.

• The credits promoted for energy access for the poor have been for consumers and for retailers, but not for the people who handle energy generation – this is particularly needed by small generators who can sell energy on a private basis to users and who require upfront capital to become established.

Box 3.8 Brazil's energy access for the poor

Brazil started restructuring its electricity sector in 1993, with completion by the early 2000s. This involved the privatization of most of its distribution and part of the generation assets and crucial national and foreign private capital investment. Despite the fact that Brazil's constitution considers electricity to be a public service under the mandate of the Federal Government (Goldemberg et al., no date), the sector reforms paid little attention to the expansion of services to low-income people.

Brazil has about 180 million inhabitants; an estimated 46 million of them live below the poverty line. In 2003, President Luis Lula da Silva inaugurated an ambitious programme – 'Fome Zero' (Zero Hunger) – aimed at fighting poverty and hunger. This programme allocates a $20 allowance per month to each Brazilian household considered to be undernourished and aims to halve the number of people living in extreme poverty by 2015 (Hicks, 2005). The objective of this programme was to reduce poverty by providing basic services and stimulating economic growth, the creation of jobs and better income for the poor. Within this programme two clear priority components were outlined, 'Zero Sede' (Zero Thirst) and 'Luz para Todos' (Light for All). The first one sought to provide drinking water and the second to supply electricity to all Brazilians by the end of 2008.

At this time access to electricity in Brazil ranged from 98 per cent in the wealthier states in the south to 82 per cent in the north-eastern states. When the programme started in 2003 the total number of people without access to electricity was 12 million, out of this 10 million were living in rural areas. The total investment levels for this project were estimated at $4.1 billion by project completion; 73 per cent of this amount was supposed to be funded by the Brazilian Federal Government and the remaining 27 per cent by regional states, government enterprises and others (Hall, 2007).

This programme was designed to provide electricity services to the poorest members of the population, more than 80 per cent of whom live in very isolated rural areas. The programme was meant to pay 100 per cent of the investment cost for 12 million people to be connected. After the service was installed each household would pay their monthly electricity bill.

The strategy considered a number of criteria including: least cost, levels of poverty and isolation, and most benefits, but the allocation of financial resources was according to the degree of poverty and access to electricity of each state. The strategy considered multi-sector participation and complementary programmes to take care of technical and social aspects, such as training people about how best to use energy and helping them to understand their responsibilities (such as paying their bills and so on). The national average of the subsidy given to the power distribution companies has been R$4,000 ($1,818) per new connection.

By the end of 2006 this programme had provided electricity to about 5 million and is planned to provide electricity to another 5.3 million during the period 2007–10.

Box 3.9 China's rural electrification

China has a population of 1.3 billion inhabitants of whom a large proportion (60 per cent) live in rural areas (Xinhua, 2006). The degree of access to electricity in China is estimated at 98.5 per cent, which is very high compared to other developing countries, particularly bearing in mind its national average per capita income of about $1,000. The high rural electrification percentage is due to the favourable policies and programmes that the Chinese Government has historically adopted in favour of energy access for rural populations. The Government of China considers energy provision to be a public service and therefore has given it continuous financial and political support. Users participate during all stages of the implementation of the programmes, but in terms of costs the local people only contribute labour.

The Chinese rural electrification took place in three stages (Program on Energy and Sustainable Development, 2006). First, from 1950 until the end of the 1970s, the driving force for rural electrification and particularly for small hydro was the development of agriculture. During this period rural electrification was slow, yet impressive and directed by strict central planning. The investments and implementation were led by the local governments and the counties. Second, from the late 1970s to the late 1990s, the Chinese Government promoted rural industrialization; during this period the central government designed strategies and policies to suit local conditions and interests, such as the strategy of 'the one who invests owns and operates'. To meet these objectives annual subsidies for the implementation of small hydropower plants and special loans were put in place. The third stage began at the turn of the century and included large-scale consolidation and upgrading of rural grids funded by a variety of sources. The driving forces were now institutional reform, boosting electricity demand, integration of the rural and non-rural electricity market, fuel substitution to contribute to environmental protection, and the improvement of quality of rural life.

The high achievements in rural electrification in China have been due to the clear role of the state in the provision of the energy electricity service, a strong institutional arrangement for rural electrification and clear roles for the different actors. A Rural Electrification Department was created within the Ministry of Water and Power from 1982–92 and was under the Ministry of Electricity from 1992–2003. This department had offices throughout the country at national, county and township levels.

The present Chinese Government plans are to have full electrification for China's rural population by 2015. For that purpose China has been implementing large-scale rural electrification programmes since the early 2000s. For example, the China Township Electrification Programme launched in 2001 by the State Development Planning Commission (now the National Development and Reform Commission) was completed in 2005. It was a scheme to provide renewable electricity to 1.3 million people (around 200,000 households) in 1,000 townships in the Chinese provinces of Gansu, Hunan, Inner Mongolia, Shaanxi, Sichuan, Yunnan, Xinjiang, Qinghai and Tibet. The programme, one of the world's largest renewable energy rural electrification programmes, used a mixture of small hydro, solar PV and wind power. This programme is now being succeeded by a similar but larger programme called the 'China Village Electrification Programme', which will bring renewable electricity to 3.5 million households in 10,000 villages by 2010. This will be followed by full rural electrification by 2015. The total cost of the programme is expected to be in the region of $5 billion. China is an example of a country adopting an explicit mandate for renewable energy for rural electrification (Jiahua et al., 2006).

Box 3.10 South Africa's energy access for the poor

South Africa has a population of nearly 48 million people of which about 75 per cent live in rural areas. At the end of the apartheid regime and the subsequent start of the democratic government of President Nelson Mandela in 1994 electricity coverage only reached 30 per cent of the South African population, with the rural areas standing at well below 10 per cent. To address this imbalance in energy access the government started the National Electrification Programme (NEP), an ambitious government programme for household electrification based on grid extension. Its implementation was the responsibility of the electricity supply industry (ESI), which is mainly in the hands of ESKOM, a government company. ESKOM holds 96 per cent of electricity generation and 100 per cent of the transmission grid, and distribution is shared between ESKOM and 235 municipal utilities (ESKOM has 60 per cent of sales and 40 per cent of customers).

The implementation of this grid extension programme was impressive; by 1999 the number of households connected to the grid was 2.7 million, which exceeded the original target of 2.5 million (SEPCO, no date). The national percentage of households with access to electricity had reached around 70.4 per cent, with 84 per cent of urban households and 50 per cent of rural households covered. As well as households this programme has also reached a large number of social services, especially schools and health centres. An important feature of this programme was the 100 per cent subsidy by the government through the electricity company.

As the implementation of this programme progressed, the government learned important lessons about costs, subsidies, viability and so on. The most important finding was that, even though the investment was entirely subsidized, the programme still required subsidies to cover operation costs, which led the government to the conclusion that for the most remote areas off-grid electricity solutions are probably a more cost-effective option. A programme of this magnitude required substantial subsidies which became a burden for the electricity company, therefore the tariffs had to be restructured. Also, new investment was needed to increase electricity generation, to avoid shortfalls in electricity supply.

The lessons learned about grid extension led the government to design and promote a new strategy for electricity services using off-grid systems for isolated households through 'the concessions model'. This model for rural electrification with off-grid systems has been matched with the 'fee-for-service' approach, in which the concessionaire provides the electricity service to the user against a monthly charge. When this model was initially designed and promoted, the aim of the government was to provide electricity services to 1.5 million households in remote rural areas. Under this rural electrification strategy the government subsidizes 80 per cent of the investment and the remaining 20 per cent is paid for by the user through tariffs.

According to the information available this off-grid programme has been mainly based on solar PV systems. By 2005 five private companies were each granted concession areas to establish non-grid energy service utilities. Limited success has been reported regarding the number of households served, as well as financial viability (Malzbender, 2005).

To meet the electricity demand in the near future, the government of South Africa has budgeted $20 billion to expand generation capacity, including the ESKOM commitment towards funding the world's largest hydro-electric power plant to be developed in the Democratic Republic of the Congo, 'the Inga Dam'.

Notes

1. For more information on the Kassala Smoke and Health Project visit the Practical Action website: http://practicalaction.org/?id=smoke_sudan
2. These costs quoted for micro- hydro, wind, solar PV and small diesel sets are from Practical Action's field experience, using equipment manufactured, designed and installed using national capacity, and operated by the villagers or communities. The cost varies depending on the intensity of use (load factor), the quality of resource, location and other factors.

CHAPTER 4
The funding gap

Estimates from the International Energy Agency (IEA, 2006) in the *World Energy Outlook* state that in order to meet the world's energy needs, a cumulative investment in energy-supply infrastructure of over $20 trillion in real terms would be needed from 2005–30. This is substantially more than the $16 trillion previously estimated in 2001.

Asian and European leaders estimate they will invest $6.3 trillion in the energy sectors of developing Asian countries by 2030. The amount represents about one-third of the total global investment in the energy sector for that period.

The International Atomic Energy Agency estimates that the investment needs of developing countries for electricity are about $160 billion per year from 2010, increasing at about 2 per cent to 3 per cent per year through to 2030. The World Bank predicts an estimated investment gap of 50 per cent, it also argues that under investment in the energy sector may cause a reduction of GDP of 1–4 per cent depending on the severity of the problem. It also predicts that the investment needs of electricity access for the poor are in the region of $34 billion per year (Lallement, 2006).

These forecasts show the enormity of the energy funding problem in developing countries and especially in the poorest economies. At the same time, the mechanisms which are supposed to target the poor are failing to deliver.

Political willingness and the commitment of governments are essential to prioritize investment in energy as key to development of the poorest sectors. Only a few governments, such as China, Brazil and Mongolia, are committed to targets and funding sources; these governments are working on a sectoral strategy of energy access for the poor, while considering issues of long-term sustainability.

On the one hand, poor countries often have neither the financial nor the technical capacity to design and implement their own policies and strategies on energy access for the poor, and their efforts are focused on the commercial and industrial sectors. On the other hand, the users or potential users, do not have the power to argue for energy needs or the knowledge to engage in debates and policy making to address their needs (they are largely 'energy illiterate' with regards to costs, sources, risks and opportunities, which complicates issues of sustainability and rational use).

Therefore the policies and strategies of the last 20 years have mainly been led by large development institutions such as the World Bank, UNDP, EU and international donors. Unfortunately the results show that progress on energy access has been slow, and in most cases the policies have not benefited the poor.

Existing commitments

International aid and energy access

According to the OECD (Organization for Economic Cooperation and Development), international aid has increased from an average of $60 billion a year in the 1990s, to over $100 billion by the mid-2000s. How much of these funds have actually been directed towards energy access is not clear, but the literature shows that energy distribution to the poor has been small and most of the investment has been directed towards support for energy security, exploration of energy resources and grid electrification (UN, 2009).

Little has been allocated to decentralized energy and energy access for the poor. UNESCO (Bequette, 1998) states:

> The World Bank spends $3.3 billion a year in the energy sector but only 7% of this goes to renewable energy in developing countries. Similarly, out of a total of $8 billion provided by multilateral aid agencies, only $1.5 billion is spent on rural electrification. Electricity supplies to the rural population are no better than 30 years ago. The reason is that maintenance and repairs are difficult, connection to the mains supply is costly, human settlements are widely scattered, access to isolated areas is difficult and consumption is often low (less than one kWh a day).

This article highlights two issues, 1) the interest in energy for developing countries and in particular for rural electrification is small, from both the financial organizations such as the World Bank and the international aid sector; and 2) energy for small isolated groups of poor people is difficult and very expensive, and the economic viability of these systems is still uncertain.

In the same article UNESCO recognizes that renewable energies have been one of the keys to sustainable development since as long ago as the 1950s, in part because of environmental concerns. However, renewable energies only started to attract more attention by the 1970s when the oil crisis became a threat to global energy security.

Nevertheless since the 1970s there have been small sums of aid money used to develop and demonstrate technologies related to efficient cooking and small-scale renewable energy, mainly aimed at helping the poor deal with their energy poverty. During the last four decades there have been a considerable number of projects related to the development and dissemination of efficient stoves and a number of devices to capture energy resources – solar, small hydro, wind and biomass – for different purposes such as drying agricultural products, pumping water for domestic purposes and agriculture, heating homes and electricity generation. These efforts have increased the numbers of people with access to energy, but unfortunately the results show that the efforts of the international aid community have been insufficient in quantity or quality to make much progress towards energy access for the poor. There are clearly gaps in technology and local capacity, which could be supported by international aid programmes.

However, it is expected that in the coming decades international aid will remain an important contributor towards energy access, in part because energy access is key to achieving the MDGs.

Governments in poor countries

Governments in developing countries are supposed to be the main drivers for providing energy access to their citizens. Ultimately they are responsible for providing access to services and creating wealth for their people. However, it is also understandable that the governments of poor countries have large and competing demands from all sectors for economic resources, and they also have national and international obligations such as paying national debt and balancing trade.

Governments in developing countries are under pressure to meet a large number of demands not only from the poor, but also from more wealthy groups, and from productive sectors. For example, industrialists ask for exemption from some taxes in order to promote competitiveness with other countries and to encourage exports, and producers for the internal market claim exemption from taxes for certain components to compete with the importation of subsidized products.

Among all these competing claims, the likelihood that the poor will gain access to government funds for energy access is generally very slim because the poor, and especially those in isolated rural areas, have little or no power, no access to policy debates, and no representation on forums discussing national policies, strategies and priorities. Therefore, they have little hope of securing changes in the allocation of economic resources for energy access.

The following two case studies from Peru and Sri Lanka demonstrate the efforts governments are making to close the energy gap between rich and poor:

Case Study 1: Government provision in Peru

Peru has a total population of 28.6 million, 72 per cent living in urban areas and 28 per cent in rural areas. By the end of 2006, only 32 per cent of the rural population had access to electricity. Although kerosene is widely used even in relatively small towns, rural areas still use wood, dung and agricultural residues. Wood is widely used in the Amazon areas where there are plentiful resources, while dung is used in the highlands of the Andes because of the scarcity of wood and agricultural residues.

In 1992 a law relating to electricity concessions was issued, this also established the basis for private participation. The activities of this sector were unbundled to cover generation, transmission and distribution; currently most generation activities are under private ownership, however transmission and a significant part of distribution are still in the hands of the government. According to government information, the interest in private participation continues. The last changes relating to private participation were enacted in 2006, including modifications to the regulations concerning tariffs. These regulations were changed in order to promote more efficient development of the electricity sector and especially the use of small decentralized energy systems. Other changes included the restructure of the Commission for the Economic Operation of the System. These changes also aimed to improve energy security, by increasing private participation and by reducing risk.

With regards to rural electrification, the government had started to implement scattered and unplanned activities in the 1960s and had continued with this approach. During the 1960s grid extensions were built in the Central Andes region. In the 1970s and 1980s activities were oriented to small hydro schemes and these so-called 'small systems' conceptualized the development of a small central generation system for a few towns in different parts of the country. By early 1990, the rural electrification coefficient had only reached 2.5 per cent. During the period 1995–2000 there was a good flow of resources for rural electrification through a combination of grid extension and the implementation of small-scale renewable systems and diesel sets and from 2000–4 rural electrification continued, but at a much slower rate than the previous years.

Despite the progress of the 1990s and early 2000s, the rural electrification coefficient in Peru remains one of the lowest in Latin America, only comparable to Bolivia and Nicaragua. Currently there are about 5.5 million inhabitants without access to electricity, with large inequalities from region to region and within towns, villages and communities in each village; the regions with greater access are located along the Pacific Ocean, due to better transport links.

Since the early 1990s the government department dealing with rural electrification in Peru has been DEP (Direccion Ejectuva de Projects) in the Ministry of Energy and Mines (MEM) and ADINELSA (Empresa Administradora de la Infraestructura Electricia S.A.), created by MEM but with limited participation by regional local governments. MEM is the promoter and policy maker, OSINERG is the regulator, ADINELSA is the administrator of infrastructure and DEP is the executor of projects.

From early 2000 the legal framework for rural electrification received significant attention and during this period several rules and regulations were enacted relating to the promotion of rural electrification, implementation, tariffs and subsidies:

- In 2001 law No. 27510 was issued to regulate tariffs with the idea of improving access and affordability for the poor through, modifying tariffs and providing cross-subsidies. It was later modified in 2004.
- In May 2002 the law for the electrification of rural and isolated areas and the country borders was issued, but has never been put to practical use.

- In August 2002 the decentralization law was issued which decentralizes decision making and resources for social development projects (although not explicitly related to rural electrification, local governments can use this law as an instrument to implement energy access projects using social development funds from government).
- In June 2005 the law for the promotion and use of renewable energy resources was approved.
- In May 2006 the rural electrification law was established.

In parallel to these laws a number of rules and regulations were issued relating to government investment and decentralization.

Currently there are two funds working in parallel: 1) The FONER, a $100 million project, was designed to implement rural electrification projects with private participation and with subsidies of up to 90 per cent. This project is based on a number of criteria, but the most important relates to co-funding and the quantity of energy used (above 100 kWh monthly) (Guerra-Garcia, 2007); and 2) The Rural Electrification Programme, which will subsidize 100 per cent of the investment and will be implemented under a different set of criteria. These programmes will have funds of around $100 million per year and will be executed entirely by DEP.

It is expected that as a result of these projects, by the end of 2010, there will be a substantial increase in rural electrification access, reducing the number of people without access to 4.4 million, and by 2015 coverage of 90 per cent of the rural population is expected. However, there are important questions which need to be answered to make efficient use of the financial resources available. First, the strategy is excessively centralized and paternalistic. DEP is the final arbiter of the project making it difficult for any other institution to implement projects, yet DEP is also the main government implementing agency.

Secondly, regional and local governments have neither human resources nor infrastructure to implement rural electrification projects. DEP is doing little or nothing to change this situation in the medium or short term. DEP has been implementing rural electrification projects within a centralized framework, and has neither been able to create capacity within the region, nor establish a delivery mechanism.

Furthermore, two of the criteria for project prioritization contradict each other. One of them requires the projects to serve the poorest, the other requires the lowest implementation costs. These two criteria are generally incompatible in rural electrification because the more expensive projects (hence, those requiring more subsidies) are those in the most isolated locations where people are more marginalized and are generally the poorest. In Peru the villages located deep within the Amazon region and in the Andes are the poorest and have fewest services, yet the implementation costs there will be the most expensive in terms of unit cost.

The infrastructure, once installed, has to be passed on to ADINELSA which owns the systems but does not operate them. ADINELSA will need ongoing resources for administration costs and for contracting the utilities to run the systems. The practical conclusion would be to assign the systems to the utilities directly, once commissioned.

An even more efficient model is from Chile, in which all rural electrification projects are executed by the utilities, and priority is given to the lowest cost. Alternatively, under the Thai model priority is given to the poor, but leaves open possibilities for co-financing, with parallel implementation.

The current regulations are clear on how small private and community-owned schemes can access subsidies. The experience of Practical Action demonstrates that small private businesses can mobilize capital effectively and maximize the benefits from existing resources. However, current regulations have not considered the connection of small schemes to the grid; there are no incentives, even though such arrangements could be feasible with small hydro schemes and utilizing private investment.

Finally, several small systems implemented by communities, municipalities and private operators are at risk of falling into disrepair due to low income from tariffs. They are not able to participate in the subsidies, because they do not fall under the current regulations.

Regarding energy for cooking, kerosene is subsidized but LPG is not. Access to cooking fuels does not present a great problem, except in the high Andes where agricultural residues and forests are scarce. During the last 20 years there have seldom been discussions about the need for policy change in this field, which gives the impression that there are no serious problems, at least in rural areas. In urban areas, where there is no free access to wood and other biomass fuels, the poor cannot afford efficient equipment, so they mainly buy cheap and inefficient kerosene cookers.

Case Study 2: Government provision in Sri Lanka

Sri Lanka has a population of 19.6 million people, 72.2 per cent of them living in rural areas; the total area of the country is 170 km^2. Seventy-five per cent of the population has access to electricity (Ramani, 2007); 4.9 million people have no access to electricity, the great majority of them living in rural areas. For cooking, 83.9 per cent of the population use firewood, 10.8 per cent use kerosene and only 4.4 per cent use LPG; agricultural residues play an important role in domestic cooking in Sri Lanka.

Average electricity consumption in the domestic sector is very low (on average 804 kWh per year per family), the main source of energy is the grid which supplies 98 per cent of houses with electricity, while the other 2 per cent is supplied by small off-grid electricity generation schemes.

In Sri Lanka electricity is under the Ministry of Power and Energy; Petroleum exploitation is under the Ministry for the Exploitation of Petroleum

Resources; and the exploration for petroleum is under another Ministry (the Ministry for the Exploration of Petroleum Resources).

Sri Lanka imports 100 per cent of all oil products; its indigenous resources are biomass, solar and wind energy, with only a small quantity of biomass being converted to charcoal and to electricity. From 1895, at the beginning of the electricity era, electricity generation was managed by private suppliers, however, in 1927 electricity came under the control of the government and it remains so to the present day. In Sri Lanka the Ceylon Electricity Board (CEB) owns and runs the electricity infrastructure. In 1983 the government created the Lanka Electricity Company Ltd (LECO), which is allowed to supply electricity and share capital with CEB. In 1997 Sri Lanka introduced directives advocating policy reform and in 1998 the first independent power producer commissioned a 51 MW system. The sector reforms enacted in 2002 provide regulations for the unbundling of the sector in generation, transmission, distribution and supply of electricity and the take over of CEB and LECO.

The main goal of the sector is to provide energy security at all times and at the least economic, social and environmental cost. The reform considers private participation to enhance sector development and improve efficiency at all stages of generation, transmission, distribution, and supply, and to extend the services to more users. In 2002 the government established the Public Utilities Commission of Sri Lanka (PUCSL), which is a multi-sector regulator.

Petroleum activities have also been dealt with by the government since the early 1960s when petroleum companies were nationalized. Yet, it wasn't until the 1980s some activities were handed over to private companies.

Both electricity and petroleum have a policy of subsidies to the poorer sections of the population. For electricity there is a cross subsidy of up to 62 per cent for domestic consumers and nearly 70 per cent for consumers categorized as religious institutions, while the industrial sectors pay an overcharge of 10 per cent and the commercial sector an overcharge of more than 50 per cent.

Although the electricity sector has progressed substantially over the last decade, this progress has been mainly through grid extension. However, the growth of energy generation has not been sufficient and consequently there are currently significant shortages of electricity, which means that many commercial institutions and small industries produce their own energy through small diesel generators.

During the period 1983–2003 more than 304,000 rural houses gained access to electricity, mainly from aid-funded projects costing a total of nearly $165 million. The last important rural electrification project started in 1997 with a planned $53 million: $5.9 million from the Global Environment Facility (GEF), $24.2 million from IDA and the balance from the private sector. This project channels money to projects under three categories: 1) energy and service delivery credit programmes for the installation of grid and off-grid hydro schemes, solar home systems and wind energy systems; 2) a pilot grid connected to wind farms; and 3) capacity building.

The government has ambitious targets of providing access to electricity to 86 per cent of the Sri Lankan population by 2010 and 95 per cent by 2015, however, clear plans and funding have still not been formulated. The main strategy is to provide more investment on grid extension in order to achieve a composition of 87 per cent supply from the grid and 8 per cent from off-grid systems.

A number of changes are also expected relating to tariff rationalization: debt restructuring and establishing an independent mechanism to evaluate the costs and efficiency of the CEB; targeting subsidies to the poorest sectors of the population; improvement of the transmission grid; improvement in demand-side and supply-side management and energy efficiency.

It is apparent that although Sri Lanka is making efforts to reach the energy poor by cross-subsidizing consumption in both electricity and petroleum products and in electricity access, there is still a clear gap in the identification of financial resources to make further progress. Although sector reforms have already been carried out, there has been no private investment and it appears that the government has no immediate plans for privatization of the electricity sector. The government-owned CEB has a large debt of nearly $1 billion (which is probably equal to or perhaps more than the value of total electricity assets if they were privatized) and is a financial burden on the government.

There is no rural electrification strategy, apart from the implementation of the $53 million plan, which will contribute to the electrification of a few thousand more families, but will not contribute greatly to reaching the committed target. Despite the reforms, there is no appropriate legal framework, no strategy and no funding to accelerate energy access for the poor.

Private investment

During the 1990s, together with energy sector reforms, there was widespread optimism regarding the role of the private sector in achieving energy access. Organizations like the World Bank and other financial institutions actively promoted the idea of private sector intervention to achieve energy access for the poor. These agencies spread influential messages such as full cost recovery of investment and the abandoning of subsidies, among others.

Nowadays, even the most radical institutions promoting the free market accept that energy access for the poor will not come about under the terms proposed during the 1990s, and that the private sector can contribute only in a limited way and only if it can generate profits. The private sector may provide skills and business know-how when they are paid to do so, but they will hardly ever make investments unless governments and international donors provide subsidies and favourable market conditions.

Private investment in developing countries in energy infrastructure reached its peak in 1997 with an annual investment of about $60 billion, followed by an immediate decrease in the following three years. From 2000 to 2006 annual investment was between $16 and $20 billion. However, the distribution of this investment between continents has been unequal, with Latin America and Asia benefiting the most, while Africa and particularly sub-Saharan Africa benefiting the least. In fact only 8 per cent of the total investment in energy has been in sub-Saharan African countries in the period 2001–6 (PPIAF and World Bank, 2007).

Sources and amounts of development finance and foreign direct investment, a portion of which is spent on technology transfer, vary widely from region to region. Countries in sub-Saharan Africa in 1997 received an average of some $27 per capita of foreign aid and $3 per capita of foreign direct investment. By contrast, countries in Latin America and the Caribbean received $13 per capita of aid and $62 per capita of foreign direct investment. Recent initiatives to spur development progress in Africa aim to respond to these disparities and to increase foreign direct investment.

It is important to point out that the quality of governance plays an important role in the efficiency of private investment. Many poor countries have serious governance problems and this jeopardizes the effectiveness of their own investment and of international aid. Furthermore providing good governance is not easy; many countries are conscious of the need for good governance but they cannot implement it for a number of reasons.

To summarize the current situation with regard to funding:

- Financial investment in energy access for the poor is urgently needed if the MDGs are to be achieved.
- Governments in poor countries have many demands from different competing sectors and will always experience great difficulties allocating resources for energy access.
- The budget allocation is carried out by government agencies, which are more interested in national energy security than the needs of the poorest.
- There has been little involvement of the poor and the NGO sector in bringing energy access issues into policy debates.
- The private sector will only marginally contribute investment in energy for the poor, especially the traditional energy private sector, which is used to making secure profits. There is some potential for mobilizing local private sector investment, but this would require considerable support and favourable market conditions.
- Good governance is necessary to make the most of the financial resources provided for energy access.

The size of the financial gap

It is difficult to make a reliable estimate on the size of the funding gap for providing energy access to the poor, partly because the cost of energy varies from one location to another with size of demand, the cost of transport, fuels, labour, financing and other factors. However, using estimates based on the number lacking electricity in rural and urban areas and the number without access to modern cooking fuel, and estimating the different implementation costs, it is possible to arrive at a figure that gives an idea of the size of the funding gap.

There are about 1.6 billion people without electricity access (0.3 billion in urban areas and 1.3 billion in rural areas), and nearly 3 billion lacking modern energy for cooking. We can assume a connection cost of $750 per family in urban areas and $1,500 per rural family to provide electricity through small decentralized energy schemes, and also assume 50 per cent of those cooking with biomass convert to liquid fuel or gas at a cost of $300 per family and the other 50 per cent gain access to better biomass cooking devices at a cost of $150 per family. These estimated costs for electricity connections cover transactional costs as well as capacity building and other institutional costs.

Based on these estimates the total investment needed to provide electricity to all of the population currently without access to electricity is in the regions of $435 billion ($45 billion in urban areas and $390 billion in rural areas). This calculation does not take future population growth into account, because on the one hand the population in rural areas is not likely to grow substantially (UN-Energy, 2005), while on the other hand the population in urban areas will grow by approximately 1.6 billion and only part of the urban population will be among the energy poor. For cooking fuel, using the above estimates, a total investment of $135 billion will be needed to provide access in the form of liquid fuel and gas to half of the population now cooking with solid biomass, and access to cleaner and more efficient biomass cooking devices to the other half.

The above estimates arrive at the total investment requirement of $570 billion. This represents only 2.85 per cent of the total investment required for energy security by 2030. However, this amount is large compared with the actual money now available for energy access. Furthermore, there is no clear provision for it; the most important climate change financing mechanisms clearly would not cover it and the private sector would not be interested in such a poor market with so many challenges.

The financial gap will not be covered by existing mechanisms or current international aid provisions unless drastic changes are made and other more imaginative mechanisms are created. One option would be to increase the global cost of energy investment by 2.85 per cent (each investor could contribute an equivalent of 2.85 per cent of its total investment towards energy access schemes). Alternatively, a new mechanism might increase the cost of fossil fuels by, for example, $1 per barrel of oil sold. This alone could provide more than the necessary funds to fill the financial gap, since the estimated oil consumption from 2007–30 is nearly 1.3 trillion barrels of oil.

If this were to happen, then considerable care would have to be taken in the design of the procedure, including the collection and the allocation of the funding, and the implementation of programmes on the ground. It would require a rapid implementation process, with about 70 million new electricity connections per year, and about double the number of conversions to cooking fuel and better cooking devices.

CHAPTER 5
Climate change and energy for the poor

Energy and climate change

There is a clear and frightening link between energy and climate change. According to the *World Energy Assessment Overview 2004 Update* (Johansson and
Goldemberg, 2004), 82 per cent of anthropogenic carbon emissions come
from energy-related activities and the other 18 per cent derive from activities
that include agriculture, deforestation, savannah burning, forest burning, agricultural residues and other uncontrolled burning.

The poor, without access to modern energies, have not shared the benefits
of the wealth created from the intensive use of energy in the last century;
however they are the most affected by the impacts of climate change due to
greenhouse gas emissions.

It is important to clarify the issue of climate change and energy access
for the poor. There is a growing campaign to stop the use of fossil fuels and
ensure all new energy development is renewable. However, some argue this
is against the development of the poor, because the poor need fossil fuels for
their development – just as the developed world used them. The issue is where
the balance lies.

Energy poverty and climate justice

Global climate change resulting from the high concentration of greenhouse
gases in the atmosphere is primarily the result of fast technological development over the last century and modern lifestyles, both requiring intensive
energy amounts. Over the twentieth century there was a 20-fold increase in
the use of fossil fuels and a tripling of the use of traditional energy forms such
as biomass. Currently, nearly 80 per cent of the world's primary energy use is
supplied by fossil fuels, which also contributes to about 80 per cent of global
CO_2 emissions.

Such development and wealth have benefited a small part of the world's
population, according to the WEC. At present, about 20 per cent of the world's
population, slightly more than one billion people, living in industrialized
countries consume nearly 60 per cent of the total energy supply. In addition,
motor vehicle emissions are a primary source of local urban air pollution.

The huge investments expected to happen by 2030 to cope with world energy demand, if present trends remain the same, will be mostly in fossil fuels
unless dramatic changes are made worldwide in government energy policies
to move from fossil fuels to clean energy resources. In fact, even conservative

projections predict that the total energy consumption from 2007–30 will be about 1.8–2 trillion barrels of oil.

The expected investments in the energy sector show that energy security for rich countries is not likely to be threatened in the coming years despite the difficulties that present themselves from time to time, such as the scarcity of resources and increases in fuel prices. Instead the main threat is from climate change. However, the importance of delivering energy access for the poor has been questioned because of climate change, despite the fact that, as will be demonstrated in this chapter, the energy needs of the poor are small and do not substantially pose a threat to climate change. It is important, therefore, to bear in mind that while energy security for the world needs to be based on clean energy resources in order to reduce the threat of climate change, energy for the poor, even if coming 100 per cent from fossil fuels, is not a threat to the planet; it would only add a marginal quantity of CO_2 to the environment. Energy security and energy access affect climate change in different ways and therefore they require different policies.

The moral responsibility of rich countries

Anthropogenic CO_2 emissions are primarily the responsibility of rich countries; their increase has been based on the development of more technologies and more intensive use of fossil fuels to power them. The range of technologies developed and their use has gone beyond the technologies associated with basic development to include production, transformation, transport, communications, improvement in living standards, comfort and recreation, and even sophisticated weapons of mass destruction. All these technologies have meant an increase in energy consumption per capita in developed countries. This energy consumption has gone beyond a contribution to prosperity, but has also been expended on pleasure and comfort, while poor countries have had opportunities neither for development nor for pleasure or comfort. Figure 5.1 shows the per capita emissions for different countries, it shows that the greatest polluters are those with high levels of development (such as Japan, UK and Norway), and those with large hydrocarbon reserves to exploit (such as Venezuela, Mexico, Qatar and Kuwait).

It is important to recognize that the damage to the atmosphere is the responsibility of rich countries that have developed their economies by emitting huge amounts of carbon dioxide. By comparison, poor countries are not only energy poor but they also suffer the consequences of global climate change. Chapter one showed that every additional kilowatt-hour consumed by poor countries contributes significantly to an increase in the human development index (HDI), especially in countries with the lowest consumption. Therefore the developed world has a moral debt to the poor, and now it is time to help them leave the trap of underdevelopment by providing access to energy for at least basic needs. Furthermore, energy access for the poor should not be

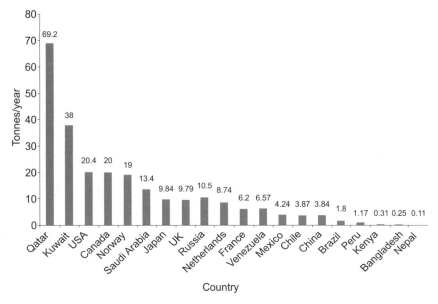

Figure 5.1 Per capita CO_2 emissions, 2004

restricted to the use of clean energy, but the poor should be allowed to choose from the full energy spectrum to fulfil their needs.

How much will energy access for the poor contribute to CO_2 emissions?

From the figures given in previous sections, meeting the estimated energy needs of the poor would do little to increase the present and future overall emissions worldwide. We can construct scenarios to estimate the sort of threat the world would face if energy access for the poor were to be achieved. This is done by estimating the total emissions that would be generated and comparing these with the total emissions currently produced worldwide, which are estimated to be 27,200 billion tonnes CO_2 for 2005 (IEA, 2006).

Scenario A

This scenario postulates that all 3 billion people without modern fuels switch from wood to LPG for cooking, and the 1.6 billion people without electricity switch from kerosene and candles to electricity for lighting and power requirements, thereby gaining the entire range of basic services and opportunities for productive uses and income generation. The total extra emissions from cooking would be approximately 305 million tons of CO_2 (equivalent to 1.1 per cent of total CO_2 emissions) from LPG. This could only be counted as additional emissions if the current supply of fuel wood for cooking was grown and collected in a sustainable way. The extra emissions from switching to electricity would amount to approximately 127 million tons of CO_2, which is an additional 0.5 per cent of emissions. Together these amount to 1.6 per cent of total global emissions.

Scenario B

This scenario postulates that only 50 per cent of those cooking with biomass get access to LPG while the other 50 per cent cooks with sustainable biomass. In addition 100 per cent of those without access to electricity gain a supply. This scenario results in an increase of about 0.55 per cent of emissions for cooking and 0.5 per cent for access to electricity. Together these amount to an increase in total emissions of 1.05 per cent.

Scenario C

In the last scenario, 50 per cent of those currently cooking with biomass switch to LPG, while the other 50 per cent cook with sustainable biomass. For electricity, 50 per cent gain access to electricity from diesel generators and the other 50 per cent from renewable sources of energy. This scenario would mean an increase in CO_2 emissions in the region of 0.8 per cent of total world emissions.

On this basis, even considering the most extreme scenario of a sudden change to modern energies, for both cooking and electricity, in both rural and urban areas, the total increase in emissions would be less than 2 per cent of current world emissions. Poor countries should therefore be granted an allowance of CO_2 emissions to provide energy access for the poor. This allowance should be targeted strictly for the poor – those currently without access to modern fuels for cooking or without access to electricity – and should not be used by the industrial and commercial sectors.

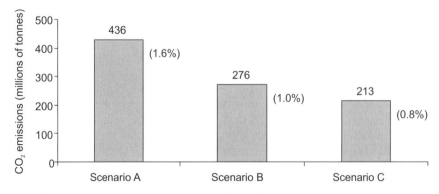

Figure 5.2 CO_2 emissions (increase compared with 2005 world total)

Energy options for the poor

An important emerging issue is the reluctance, or low level of support, from many policy makers, international aid agencies and environmentalists, to consider the full range of energy options to meet the energy needs of the poor. To deliver access to modern energy for the poor it is essential to maintain the full range of energy options including grid and decentralized electricity, renewable energy and improved cooking stoves, as well as options based on fossil fuels. Energy for the poor is not a threat to climate change even in the worst case scenario, since as shown here, even if 100 per cent of the 2 billion energy poor switched to fossil fuels global emissions would increase by only 1.6 per cent. Is it not better to use the world's most precious resources to tackle deprivation, drudgery and exposure to smoke from dirty fuels, than to drive four-by-four town cars or to heat private swimming pools?

We need to look at the most suitable energy options independently of whether or not these are renewable sources, since the provision of energy access to the poor would only marginally increase the emission of greenhouse gases. Furthermore poor countries have emitted lower levels, hence should be allowed to increase these amounts, especially for the purpose of energy access for the poor and the achievement of the MDGs.

This is not to suggest, however, that poor people in developing countries should not benefit from the increasing funds channelled into low-carbon energy technologies, or be allowed to leapfrog to new renewable technologies wherever possible.

Low-carbon funding mechanisms

Considerable global funds are now being channelled into low-carbon technologies, through donor arrangements and carbon trading. Unfortunately, the vast majority of this new funding has bypassed low-income countries, with the poorest sectors of society receiving least financial support. Funding could be directed to schemes that would reduce carbon emissions as well as promote energy access such as small-scale renewable energy schemes in rural areas as well as local management of forestry and reforestation.

The Clean Energy for Development Investment Framework (CEIF)

The CEIF comes in response to the recommendations of the G8 summit in July 2005 to create 'a new framework for clean energy and development'. The first phase, a draft proposal, was completed and submitted at the WB-IMF annual meetings in Singapore in September 2006. It included analysis of the adequacy of existing international finance institution (IFI) mechanisms to address: the long-term energy needs of developing countries; the mitigation of greenhouse gas emissions; and the adaptation needs of developing countries. More research is underway regarding longer-term programmes of country-level activities and global research.

An action plan was prepared by the World Bank and endorsed by the board in March 2007, it considers three major objectives (or pillars): 1) energy for development and access for the poor; 2) transition to a low-carbon economy; and 3) adaptation to climate change. During the preparation process this proposal received criticism from several organizations, especially from civil society organizations, who considered that the document does not show sufficient leadership from the World Bank in tackling the problem of emissions. Of particular concern is:

- The continuation of World Bank funding in the range of $2–3 billion a year on greenhouse gas producing fossil-fuel projects.
- The Bank's failure to sufficiently support energy access for the poor.
- The fact that that only 5 per cent of the Bank's investments on energy is directed to renewable energies.
- The continued support for unsustainable large-scale hydropower.
- The poor performance of GEF funds for renewable energies for development, and the poor performance of previous renewable projects with GEF support, such as the solar PV initiative launched in 1994.

Therefore, the civil society organizations urged the World Bank to eliminate subsidies on fossil fuels and redirect them to renewable energies.

In September 2007 the World Bank submitted a progress report on the action plan to the board (World Bank, 2007). This report responds to the criticisms raised and partially provides answers. In fact it highlights the issues of energy poverty and the barriers to providing energy to the poor, and it highlights the importance of energy access to meet the MDGs. However, it is based on an expectation that the private sector will play an important role in both investment and removal of barriers regarding energy access for the poor, which is probably unrealistic because it is not an attractive market for private investment (see Chapter 2). In addition, it only partly answers the criticism about the Bank's support for fossil fuels.

The Global Environment Facility

The GEF was established in 1991 to fund environmentally sound projects and programmes in developing countries. GEF supports projects in six different areas: biodiversity, climate change, the ozone layer, international waters, land degradation and persistent organic pollutants. Since 1995, the GEF strategy follows guidance from three international conventions – the Convention on Biological Diversity, the UNFCCC and the Montreal Protocol – for its biodiversity, climate change and ozone-layer areas of funding respectively (Solis, 2007).

There are three main agencies that have been implementing the GEF's climate change activities: UNDP, UNEP and the World Bank. Four regional development banks manage and execute GEF's climate change projects: the Asian Development Bank, the African Development Bank, the Inter-American Development Bank and the European Bank for Reconstruction and Development, the Food and Agricultural Organization (FAO), UN Industrial Development Organization (UNIDO) and the International Fund for Agricultural Development (IFAD).

Until 2006, the GEF portfolio had allocated nearly $7 billion to almost 2,000 projects. The share for climate change projects was $2.37 billion (Solis, 2007).

GEF's projects are funded under its three main areas: large-scale projects (above $1 million), medium-scale projects ($50,000 up to $1 million); and small projects (below $50,000). Large-scale projects are implemented by GEF's agencies, regional development banks and operational focal points in each recipient country. Medium projects are also implemented by these institutions. Other activities funded by GEF are enabling activities, which cover the identification of projects, project preparation, facilitation and the small and medium enterprise programmes aimed at encouraging the creation of enterprises.

GEF's climate change portfolio

During GEF's first decade (1991–2000) it allocated more than $1 billion to more than 270 climate change projects; 15 years later, by 2006, this amount doubled to more than $2 billion and more than 600 climate change projects, most of the funding (over 90 per cent) for the two periods has been allocated to large-scale activities.

During this period, UNDP has been involved with the largest number of projects – more than 60 per cent of the total number. However, the World Bank has implemented more expensive projects (60 per cent of the funds over the two periods). The allocation of the GEF's funds has mainly focused on renewable energy and energy-efficiency project activities. It is also important to point out that the funding for sustainable transport projects increased by 4 per cent during these two periods.

The GEF renewable energy technologies portfolio (either already implemented or being implemented) is made up of 136 projects costing a total of $851 million. Yet, only 54 of them are small renewable projects under 15 MW (from 1991–2006), representing 27 per cent of the investment ($243 million).

The above figures indicate that the number of families who benefited from small renewable energy projects funded by GEF from 1991–2006 is small, although there are no exact figures from GEF or other organizations. Assuming that it costs an average of $1,500 per family to gain access to electricity, it is estimated that the total number of families who have benefited from GEF is around 150,000 (or 750,000 people) over a period of 15 years.

The Clean Development Mechanism (CDM)

In 1997, the Kyoto Protocol of UNFCCC was signed by more than 100 countries during COP-3 in Japan. This international agreement commits developed nations to reducing greenhouse gas emissions to at least 5 per cent below 1990 levels during 2008–12 (UNFCCC, no date). The protocol includes three market-based mechanisms: emissions trading; joint implementation; and the

Clean Development Mechanism (CDM) as instruments to support countries in achieving their commitments.

The CDM is intended to assist developing countries with mitigation and adaptation measures. The idea for this fund was also proposed during COP-3. By this time, emissions trading and joint implementation mechanisms were already being introduced by the UNFCCC. The CDM has two main objectives: 1) to reduce emissions; and 2) to support sustainable development within developing countries. An agreement stating that the proceeds from the fees associated with CDM activities will be used to cover administrative costs and to assist vulnerable developing countries in adaptation measures was reached. This agreement and the second CDM objective – sustainable development support – distinguishes the CDM from the other two mechanisms (Grubb, 2003).

So far CDM has shown poor performance in relation to energy for development and even poorer results in tackling energy access for the poor. Most of the CDM funds have been used in 'transition economies'; by June 2008, out of a total of 1,197 registered projects, 30.08 per cent were in India, 24.14 per cent in China, 12.2 per cent in Brazil, 8.94 per cent in Mexico, 2.76 per cent in Malaysia, 2.12 per cent in Chile, and only 19.72 per cent in the rest of the world. Out of the 2.12 per cent of funds spent in Africa most were for projects in South Africa (CDM, undated).

The main barriers for small-scale energy access projects are the rigidity of the rules and transaction costs, which are designed for large projects with millions of tons of CO_2 savings, while for small projects, like those needed for energy access for the poor and specifically for electricity generation in rural areas the results are negligible.

For this reason, there was an urgent need to convince the UNFCCC and the signatory countries to change the rules, making them more flexible, especially regarding monitoring and evaluation, registration procedures and verification procedures, and thus to be able to create new projects to contribute to energy access for the poor.

Small-scale CDM projects. Small-scale CDM projects are defined under three main types (World Bank, 2003b):

1) Projects that use renewable resources such as solar, wind, hydro, geothermal, wave and tidal power; these are renewable energy projects with a maximum output of installed capacity of 15 MW.
2) Energy efficiency improvement project activities which reduce energy consumption, on the supply and/or demand side, by up to the equivalent of 15 GW-hour per year.
3) Other project activities that reduce emissions from sources that emit less than 8 kt CO_2 equivalent per year; examples include non-electricity projects such as agriculture, switching from fossil fuels, transport, methane recovery and emission avoidance.

Transaction costs. Undertaking a CDM project poses many risks and significant transaction costs for the project developer due to the complexity of the CDM accreditation process itself, and the costs of accounting for Certified Emissions Reductions (CER). In the context of the Kyoto procedures, transaction costs arise from the administrative process, which depends on the institutional framework. Within the CDM project cycle transaction costs occurred at every stage.

Under market conditions, small projects are the most vulnerable as they are exposed to transaction costs which sometimes can be even higher than the proposed activity costs. Studies suggest that transaction costs for large projects can be as high as $270,000 (Solis, 2007). For small projects, even considering the simplified methods and procedures, the transaction costs are projected to be between $23–$78,000 (IETA, 2004). Therefore there has been limited uptake of CDM by small projects.

Bundling. This concept, which involves putting together a number of different projects for funding, was introduced to reduce overall transaction costs of CDM small projects. However, there is now sufficient evidence that bundling can only reduce costs modestly and hence only modestly increase financial viability. Bundling has had little impact, for example, on the low level of CDM funding for Africa. As a result many are urging for changes in order to make CDM more accessible and useful.

The main proposal is the creation of a sub-category for very small projects (Solis, 2007). This sub-category might attract the interest of investors in projects in rural areas such as off-grid energy plants. Practical Action's experience has been that flexibility and capacity building can help in overcoming these challenges and lowering the costs, so proposes greater flexibility in the rules for base line studies, evaluation procedures, certification, and monitoring of small-scale projects if some tangible results are to be expected in the near future.

To summarize findings related to the low-carbon funding mechanisms:

- None of the existing global mechanisms for renewable energies (GEF, CDM or joint implementation) have contributed greatly to energy access for the poor.
- The approach and concepts considered under the Clean Energy Framework (market approach) are unlikely to produce significant funds for energy access.
- There is concern by many organizations about the effectiveness of GEF for energy access for the poor and about the low performance of CDM in energy for development.
- Only $234 million has been invested in small-scale projects in the whole period of implementation of the GEF (from 1991–2006). This amount is a small part of the $7 billion invested by GEF, with 35 per

cent of its total allocation of funds having gone to climate change projects.

- The transaction costs under CDM have been a very significant barrier for small-scale projects.
- Bundling has been considered a way to overcome large transaction costs, however there is still a lot to learn and it has not yet proven to be as effective as was expected.

CHAPTER 6
Conclusions and recommendations

Energy poverty in developing countries is severe; there are 1.6 billion people without access to electricity and more than 3 billion cooking with solid fuels – biomass and coal – under unhealthy conditions. The impact of energy poverty affects primarily women and children, many of whom have limited access to adequate social services such as health, education, and communications. Furthermore, they do not have opportunities to increase their production and employment due to a lack of access to electricity. They are faced with the drudgery of collecting wood and cooking fuels, and their health is affected by smoke from the kitchen and by low-quality or non-existent health and other basic services.

Despite the huge energy investment forecast for the next 25 years – about $20 trillion to be invested by 2030 of which about $8 trillion is to be invested in developing countries – even optimistic forecasts estimate that about 1.4 billion people will still remain without access to electricity.

The most important organizations related to human development and finance (UN, World Bank, EC) agree that energy access for the poor needs to be in place first if the MDGs are to be achieved. However, the UN recognizes eight years after the Johannesburg Summit, when the MDGs were validated, most poor countries are off-target regarding the achievement of the MDGs. The poorest countries, such as those in sub-Saharan Africa have progressed least, both with less progress on energy access and fewer achievements with regard to the MDGs.

Subsidies for the commercialization of energy supplies generally benefit the middle classes more than poor people. The poor who have no access to any energy services cannot benefit from any kind of subsidy; in rural areas there are only a few who have access to electricity, gas and kerosene and who benefit from subsidies for these products.

However, a number of energy options could be developed for the poor, varying according to the context. The urban poor generally gain access to electricity and modern fuels for cooking much earlier than rural people, but the main problem that the urban poor face is affordability. In rural areas the main option for cooking fuel for the foreseeable future appears to be biomass, while for electricity the more appropriate energy options are usually small decentralized systems, especially those based on renewable energies.

There are significant barriers, however, that prevent the poor from receiving affordable, reliable and sustainable energy. The biggest barrier of all is poverty, and poverty and a lack of energy access mutually reinforce each other.

The energy reforms of the 1980s and 1990s, which might have helped over-come these barriers, have not contributed to energy access for the poor; in fact in some cases the opposite has occurred and their access to energy has reduced. This is because energy reforms have been carried out with the com-mercial energy sectors, and not the poor, in mind.

National governments of poor countries do not have specific policies and strategies for energy access; neither have they had the institutional arrange-ments to achieve it. Both financial resources and political willingness on the part of national governments and aid agencies are at present lacking, but needed if the poor are to gain access to modern energy supplies. As chapter four described, according to estimates from the World Bank, the investment gap regarding energy in developing countries is around 50 per cent of that needed for the period to 2030. There is no doubt that that gap will most affect the poorest.

The 'free market approach' is the strategy being promoted by large financial organizations, and commonly agreed by development agencies and aid agen-cies regarding energy investment in developing countries. However, while the free market approach may work well in the commercial and productive sectors of the economy it has largely failed to bring about energy access for the poor.

The existing financial mechanisms expected to contribute significantly to energy access for the poor are failing dramatically. In the past 15 years the GEF has invested only a small amount of money in energy systems below 15 MW (at least some of them were for isolated energy schemes). The CDM is not con-tributing to energy for the poor; small amount of investments are channelled into energy security, mainly in 'transition' countries.

Moreover, other sources of international aid have failed to reach the energy poor. The UK Commission for Africa paper, examining the continent with least energy access for the poor, focused on grid interconnection and intercity roads (national and international), but paid little or no attention to access for the poor.

Recommendations for promoting energy for the poor

The strategies of development agencies, international aid agencies and the governments of developing countries, should focus on energy access for the poor as a global issue, one which has great potential for contributing to the attainment of the MDGs.

Recommendations for policy

Bring energy access for the poor to the top of the international development agenda. Despite the widespread acceptance that energy is critical for development among multilateral and bilateral agencies, governments, academia and civil society, energy is not a high priority issue in policy debate. There are no specif-ic objectives or targets within the MDGs on energy access. Energy is a critical issue for human development and it should therefore be considered a human right, and should be provided on the basis of justice for the poor.

The funding gap must be acknowledged by the international community as well as by governments. This could lead to finding global solutions through cooperation with the developed world and better funding for new investments in poor countries.

Political will from government. Important change is only possible with political support from the highest levels of government. The free market approach alone will not bring about energy access for the poor; governments need to take responsibility.

Clear and specific policies and strategies and good governance. The experience of the last three decades shows that neither global strategies nor country strategies have been clear enough to tackle energy access. The global strategies led by World Bank and UNDP have changed several times and have demonstrated slow progress. The governments of poor countries have rarely had specific policies regarding energy access for the poor; instead they have followed the lead of the multilateral agencies.

Sustainability of energy access. This is an issue not yet resolved, but to make progress towards the long-term provision of energy supplies the following activities need to be widely promoted:
- Building national and local capacity as the most effective way to contribute to affordability, accessibility and sustainability.
- Mobilization of local capital can contribute to energy access for the poor.
- Energy literacy is important to contribute to the long-term viability of the systems and to better relations between the providers and users of energy.

Recommendations on greenhouse gases and energy access for the poor

An allowance of CO_2. It has been demonstrated that if the poor gained access to modern fuels, even under the most CO_2 emitting scenarios of energy supply consisting entirely of fossil fuels for electrification and for cooking, the total increase in carbon emissions would be as little as 2 per cent of current world emissions. Developing countries should therefore have an allowance of 2 per cent to provide energy access for the poor. This allowance should be divided between countries according to the energy needs of the poor in these countries, and excluding any other energy needs (for example, transport, commerce, industry, mining, etc.) The full range of energy options should be made available and the energy mix should not be subjected to screening with regards to climate change, but consider instead criteria of sustainability, cost and availability.

Recommendations on financing the funding gap

The funding gap for energy access for the poor. It is clear that there is a huge funding gap, which cannot be covered by the poor themselves. The initial investment of about $435 billion is needed to provide electricity to 100 per cent of the current population. In addition, an estimated investment of $135 billion could help to switch about 50 per cent of the population cooking with solid biomass and could provide access to efficient and clean cooking facilities for those remaining on solid biomass. This has to be funded by international aid, climate change financial mechanisms and by governments.

Alternative climate change mechanisms. Energy access for the poor should be financed by directing resources from the CEIF, CDM and GEF to the countries which most need support. Currently these resources are not used for the poor but for the growth of economies in transition countries such as China, Brazil, South Africa and Mexico.

Finally, a new mechanism must be developed which can transfer increasing amounts of the growing carbon market fund to accessing energy projects and away from large industrial projects.

Annex 1: Practical Action's experience with energy access for the poor

Energy technology development and transfer

Practical Action has developed, tested and promoted a range of technologies in small-scale decentralized energy generation, micro hydropower, wind energy, biogas generation, efficient cooking stoves, and has recently begun researching opportunities for using bio-diesel for energy supplies for the poor. The main objective behind this work has been cost reduction, simplicity (ease of operation and maintenance) and reliability.

Micro hydro

This is the most mature energy generation technology developed and promoted by Practical Action, who have worked in this field since the late 1970s. Practical Action's country offices have been actively involved in this technology – in particular in Peru, Sri Lanka and Nepal, where several hundred schemes are operating successfully. In recent years there have also been some activities in Zimbabwe and Kenya.

Practical Action has developed, tested and transferred technologies for harnessing small hydro energy resources, ranging from a power size of a few hundred watts to 500 kW. Practical Action has developed different turbine designs for a range of conditions, from high-altitude hilly country to low-lying rivers, which comply with the requirements of the different sites and produce a range of power sizes. This has allowed micro hydro technology to be feasible in a far greater number of locations.

Wind energy

Wind energy is not new; during the early 1980s Practical Action successfully developed and transferred a windmill for water pumping. However, it has not pursued this technology, and at present there is only one manufacturer of this particular water pump in Kenya.

Wind electricity generation is a new area of work which started in 1998 and was funded by DFID. Three Practical Action offices took part in the development of this technology: Peru, Sri Lanka and the UK. The results of this project were very encouraging: two models producing 100 W in Peru and 200 W in Sri Lanka were successfully designed and tested. In each of the countries the technology has been transferred to small private companies who are now commercially producing systems. In addition, the Sri Lanka office has managed to transfer the technology to Nepal.

Cooking stoves

More efficient cooking stoves is another area Practical Action has been involved in for several years. The offices more active in developing this technology are Kenya, Sri Lanka, Nepal and Bangladesh. This work has resulted in a range of technologies to cope with different fuels: wood, agricultural residues, twigs, dung and others. A few of the models developed by Practical Action are now made commercially and sold on the market.

The main reasons for the development and promotion of stoves by Practical Action are the inefficiency of traditional stoves, the drudgery involved in the collection of wood or other fuels, and the high levels of smoke pollution which presents a health threat to women and children.

One of Practical Action's most successful projects has been the design and promotion of the Anagi cooking stove, which was first developed and tested in Sri Lanka in the 1980s and since then has spread to several other countries in Asia. The Anagi is a two-pot cooker made of clay; among its main characteristics are its portability and low cost.

Solar energy

Recently Practical Action has gained experience in the design and installation of hybrid wind solar PV systems. Practical Action Sri Lanka has installed a few systems of this kind and the Peru office is currently installing a total of 20 hybrid systems for the Ministry of Energy and Mines. Hybrid systems of this type represent a positive step forward and contribute to the sustainability of the system since the security of the energy supply is an important issue. Generally wind and sunshine complement each other well during the calendar year: during the summer there is generally less wind but more sunshine, while the opposite is the case during winter.

Biogas plants

Biogas has not been one of Practical Action's strategic technologies, however, there have been important opportunities for intervention, such as in Sri Lanka and Nepal, where biogas plants previously installed by other institutions have been reviewed and redesigned.

Energy-efficient brick making

During the 1990s Practical Action worked on the development and promotion of fuel-efficient kilns for informal sector brick making. Practical Action developed projects and disseminated technology in Zimbabwe, Sudan and Peru. All these projects were successfully completed.

LPG for cooking

In the last three years Practical Action has started to work on fuel switching with poor communities in Sudan. This work has not involved technology development, but the promotion of the fuel. However, this is an area requiring further work not only for the promotion of the technology but also for setting the standards.

Bio-diesel

The Peru office started to promote bio-diesel in 2003, with the support of the National Research Council (CONCYTEC). Some progress is already apparent, with bio-diesel working well technically for small-scale electricity generation. Practical Action is now one of the leaders in the field of bio-diesel in Peru. However, there is still a long way to go before the case is proven regarding whether or not bio-diesel is viable socially and economically as a source of energy for the poor.

In addition to these technologies, Practical Action has been involved in the development or implementation of other technologies related to energy, such as handpumps (Peru and Zimbabwe), energy efficiency in small informal metal-casting facilities (Peru) and drying systems for agricultural products.

Practical Action's working approaches

Practical Action has been active in promoting sustainable energy systems and enhancing the positive impacts energy use. The approach promoted throughout the country offices involves: community participation in the evaluation of energy options and resources, planning, design, implementation and running of the systems. Practical Action is active in the implementation and promotion of participative approaches and has included cross-cutting issues in its projects, such as gender inclusion, training, and income generation activities.

Financial issues

Although most of Practical Action's energy access projects have been funded by grants, the organization has also been active in designing and promoting financial schemes, such as loans or revolving funds. In Peru Practical Action has been operating the Revolving Fund for Micro Hydro Schemes since 1994, and in Sri Lanka Practical Action played an important role from 2000–1 in the design and implementation of a revolving fund for renewable energies that is still in operation under the management of the World Bank.

In Sudan and Kenya Practical Action has been implementing credit systems for access to cleaner cooking devices and fuels. In Kenya this has funded fuel wood stoves and in Sudan the switch from charcoal and wood to LPG.

Information dissemination

The energy teams of Practical Action have provided information to a wide audience through leaflets, case studies, impact evaluations, surveys, and the web pages of different country offices. The country offices and Practical Action Publishing have also produced a long list of energy publications related to technology, working approaches, impact studies, financial issues and policy issues.

Supporting small energy enterprises

Practical Action has also been active in technology transfer to small local enterprises and in building national capacity in developing countries to assess energy needs, and to design and implement systems. In Peru, Sri Lanka and Nepal there are currently a number of small manufacturers and consultants in the field of renewable energies as a result of these interventions.

Policy research

Over the last five years Practical Action has been engaged in policy research activities, through projects such as the Sustainable Energy Options for rural communities in Latin America, SPARKNET, Partners for Africa, and African Voices.

Training and capacity building

Practical Action has developed a number of international courses in Asia, Latin America and Africa. Currently it is running a training course on renewable energies once a year in Cajamarca Peru.

Advocacy

Practical Action has contributed to global debates on smoke pollution, and is currently engaged in debates on the wider issue of energy access for the poor. During 2006 and 2007 Practical Action began lobbying policy makers and governments to make policy changes in favour of the poor within the framework of international events, such as the Commission for Sustainable Development (CSD-14 and CSD-15), and the G8 Summit in Gleneagles in 2005.

References

ADB (Asian Development Bank) (2005) 'Rural Electrification Republic of the Fiji Islands Project', Technical Assistance Consultant's Report, New York, available from: http://www.adb.org/Documents/Reports/Consultant/35487-FIJ/35487-FIJ-TACR.pdf [accessed 1 January 2010].

AEPC (Alternative Energy Promotion Centre) (2006) *Subsidy for Renewable (Rural) Energy*, Government of Nepal, Ministry of Energy Science and Technology of Nepal, available from: http://aepc.gov.np/downloads/SubsidyPolicyEnglish2006.pdf [accessed 20 January 2010].

Ahammed, F. and Ahmed Taufiq, D.A. (2008) 'Case study: application of solar PV on rural development in Bangladesh', *Journal of Rural and Community Development School of Business Studies* 3: 93–103.

Amerasekera, R.M. (2006) 'Commercialised stove production in Sri Lanka: 30,0000 stoves a year – a success story', INFORCE, available from: http://www.inforse.org/asia/pdf/Anagi_commercial.pdf [accessed 20 January 2010].

Amarasekara, R.M. and Atukorala, K. (2003) 'Historical timeline for subsidy to commercialization of improved cookstoves: path leading to sustainable stove development and commercialization in Sri Lanka', IDEA, Kandy, Sri Lanka.

Barefoot College, India (2003) 'Solar energy to meet basic human needs in the Himalayas', Ashden Awards for Sustainable Energy, available from: http://www.ashdenawards.org/files/reports/Barefoot%20college2003%20Technical%20report.pdf [20 January 2010].

Barnes, D. (ed) (2007) *The Challenge of Rural Electrification: Strategies for Developing Countries*, RFF and ESMAP, Washington D.C.

Barnes, D. and Foley, G. (2004) *Rural Electrification in the Developing World: A Summary of Lessons from Successful Programs*, UNDP/ESMAP and World Bank, Washington D.C.

Barnes D., Douglas, F. and Willem, F. (1997) 'Tackling the rural energy problem in developing countries', *Finance and Development*, June 1997, pp. 11–15.

Barnes, D., Openshaw, K., Smith, K.R. and van der Plas, R. (1994) 'What makes people cook with improved biomass stoves?, *World Bank Technical Paper No 242, Energy Series*, The World bank, Washington D.C.

Beck, F. and Martinot, E. (2004) 'Renewable energy policies and barriers', Academic Press, Elsevier Science, available from: http://www.martinot.info/Beck_Martinot_AP.pdf [accessed 1 January 2010].

Benallou, A. (no date) 'Sun Light Power Moroc (SPM)', Rabat, Morocco, available from: http://resum.ises.org/documents/SunlightPowerMaroc.pdf [accessed 1 January 2010].

Bequette, F. (1998) 'Renewable energy: winds of change' [online], *UNESCO Courier*, available from: http://findarticles.com/p/articles/mi_m1310/is_1998_May/ai_20825358/ [accessed 20 January 2010].

Boyd, E. (2006) 'Biogas support programme (BSP) – Nepal hits 150,000 household biogas plants' [online], HENDON Household Energy Network, available from: http://www.hedon.info/goto.php/726/news.htm [accessed 1 January 2010].

BREDL (Blue Ridge Environmental Defense League) (2002) 'Waste gasification impacts on the environment and public health', *A Blue Ridge Environmental Defense League Report*, North Carolina, USA.

Brew-Hammond, A. and Crole-Rees A. (2004) *Reducing Rural Poverty through Increased Access to Energy Services: A Review of the Multifunctional Platform Project in Mali*, UNDP, New York.

Cabraal, A., Cosgrove Davies, M. and Schaeffer, L. (1996) ' Best practices for photovoltaic household electrification programs: Lessons from experiences in selected countries', *World Bank Technical Paper Number 324*, Asia Technical Department Series, The World Bank Washington, D.C.

CDM (undated) 'CDM statistics' [online], available from: http://cdm.unfccc.int/Statistics/index.html [accessed 1 January 2010].

CER-UNI (1998) 'Estudio nacional sobre energías renovables en el Peru', case study from the National University of Engineering, Lima, Peru.

Collier, Paul (2007) *The Bottom Billion: Why the poorest countries are failing and what can be done about it*, Oxford University Press.

Columbia Business School Alumni (2004) 'United Nations expert debunks myths' [online], *Alumni News*, Columbia University, USA, available from: http://www.gsb.columbia.edu/cfmx/web/alumni/news/article.cfm?legacy=0406/mcdade [accessed 1 January 2010].

Cozzi, L. (2006) *Energy and CO2 Emissions Trends in the Transport Sector*, Stern Review Team, London, available from: http://www.hm-treasury.gov.uk/media/4/C/stern_transportseminar_cozzi.pdf [accessed 1 January 2010].

Croxford, B. and Rizig, M. (2007) 'Is photovoltaic power a cost-effective energy solution for rural people in western Sudan?', Bartlett School of Graduate Studies, University College, London, available from: http://eprints.ucl.ac.uk/2640/1/2640.pdf [accessed 1 January 2010].

Cust, J., Singh, A. and Neuhoff, K. (2007) 'Rural electrification in India: economic and institutional aspects of renewables', Faculty of Economics, University of Cambridge, available from: http://www.electricitypolicy.org.uk/pubs/wp/eprg0730.pdf [accessed 1 January 2010].

DFID (Department for International Development) (2002) *Energy for the Poor, Underpinning the Millennium Development Goals*, DFID, London.

Edwards, R.D., Smitha, K.R., Zhangb, J. and Mac, Y. (2004) 'Implications of changes in household stoves and fuel in China', *Energy Policy* 32: 395–411.

ESMAP (2000) *Energy Services for the World's Poor, Energy and Development Report*, International Bank for Reconstruction and Development, Washington D.C.

European Commission (2005) 'A vision for photovoltaic technology', Report by PV-TRAC, European Commission, Brussels, available from: http://ec.europa.eu/research/energy/pdf/vision-report-final.pdf [accessed 1 January 2010].

European Commission (2008) 'Evaluation of EC support to partner countries in the area of energy', *Evaluation to the EC Final Report*, available from:

http://ec.europa.eu/europeaid/how/evaluation/evaluation_reports/reports/2008/1192_vol1_en.pdf [accessed 20 January 2010].

GNESD (Global Network on Energy for Sustainable development) (2004) 'Energy access theme results', in S. Karekezi and A. R. Sihag (eds.), *Synthesis/Compilation Report*, available from: http://www.gnesd.org/Downloadables/Energy_Access_I/Synthesis_Report_ver_30_April_2004.pdf [accessed 20 January 2010].

Goldemberg, J., Johansson, T., Reddy, A. and Williams, A. (2004) 'A Global Clean cooking fuel Initiative', *Energy for Sustainable Development*, 3(3): 5–12.

Goldemberg, J., La Rovere, E.L. and Coelho, S.T. (no date) 'Expanding access to electricity in Brazil', AFRPREN /FWD, available from: http://www.afrepren.org/project/gnesd/esdsi/brazil.pdf [accessed 1 January 2010].

Grubb, M. (2003) 'The economics of Kyoto Protocal', *World Economic* 4(3): 143–189.

Guerra-Garcia, G. (2007) 'Diagnóstico y propuesta de agenda y políticas para la promoción del acceso a energía en zonas rurales del Perú', *Policy Research Report Document*, Practical Action.

Gül, T. (2004) 'Integrated analysis of hybrid systems for rural electrification in developing countries', *Master Thesis*, Department of Energy Processes, Royal Institute of Technology, Stockholm.

Hall, D. (2007) 'Public sector finance for investment in infrastructure – some recent developments', Public Services International Research Unit, the University of Greenwich, UK, available from: http://www.psiru.org/reports/2007-04-U-pubinv.doc [accessed 1 January 2010].

Hicks, K. (2005) 'The *Fome Zero* program – Brazil's losing struggle to help the hungry: Lula's leadership fading' [online], COHA Research Council, available from: http://www.coha.org/the-fome-zero-program—losing-struggle-to-help-the-hungry-lula%E2%80%99s-leadership-fading/ [accessed 1 January 2010].

Hood, A.H. (2006) 'Project proposal on cleaner and healthier cooking in Al Fasher', Practical Action, Rugby, UK.

IEA (International Energy Agency) (2006) 'Electricity access', *World Energy Outlook 2006*, Paris, available from: http://www.worldenergyoutlook.org/docs/weo2006/Electricity.pdf [accessed 1 January 2010].

IETA (International Emissions Trading Association) 'Estimating the Market potential for clean development mechanism: review of models and lessons learned', *PFCplus Report*, Washington D.C., available from: http://www.iea.org/papers/2004/cdm.pdf [accessed 20 January 2010].

INFORSE-Asia (2006) *Sustainable Energy Solutions to Reduce Poverty in South Asia – A Manual* [online], The International Network for Sustainable Energies, available from: http://www.inforse.org//asia/M_III_stoves.htm [accessed 1 January 2010].

Jiahua, P., Meng, Li., Xiangyang, W., Lishuang, W., Elias, R.J., Victor, D.G., Zerriffi, H., Zhang, C. and Wuyuan, P. (2006) 'Rural electrification in China 1950–2004: Historical processes and key driving forces', *Working Paper 60*, Program on Energy and Sustainable Development, Standford University.

Johansson, T. and Goldemberg, J. (eds.) (2004) *World Energy Assessment Overview: 2004 Update*, UNDP, UN-DESA and the World Energy Council.

Kalumiana, O. (2004) 'Energy services for the urban poor in Zambia;, *Working Paper No. 318*, AFREPREN/FWD, Nairobi.

Kaufman, S. with Duke, R., Hansen, R., Rogers, J., Schwartz, R., and Trexler, M. (2000) 'Rural electrification with solar energy as climate protection strategy', *Research Report No9*, REPP (Renewable Energy Policy Project), available from: http://www.repp.org/repp_pubs/pdf/rurElec.pdf [accessed 20 January 2010].

Koharznovich, V. (2008) 'Research commercialization – The role of international organizations', presentation from the International Conference on Investing in Innovation, 10–11 April 2008, Geneva, UNIDO, available from: http://www.unece.org/ceci/ppt_presentations/2008/fid/Vladimir%20Kozharnovich.pdf [accessed 1 January 2010].

Lallement, D. (2006) 'Energy financing issues, prospects for developing economies', IAEA Programmatic Meeting, Viena, 31 October, 2006, available from: http://tc.iaea.org/tcweb/abouttc/strategy/Thematic/pdf/presentations/energysystemplanning/Energy_Financing_Issues-Prospects_Devp_Econ.pdf [accessed 1 January 2010].

Lapping, M. B. (2005) 'Rural policy and planning', *Rural Development Paper 31*, Northeast Regional Centre for Rural development, Pennsylvania State University, available from: http://www.nercrd.psu.edu/Publications/rdppapers/rdp31.pdf [accessed 1 January 2010].

Long, S. (2006) 'Approaches to using renewable energy in China', School of Photovoltaic & and Renewable Energy Engineering, The University of South Wales, Sidney, available from: http://www.ceem.unsw.edu.au/content/userDocs/LongSeng_To_000.pdf [accessed 20 January 2010].

Luque, A. and Hegedus, S. (eds.) (2003) *Handbook of Photovoltaic Science and Engineering*, Wiley and Sons, Hoboken, NJ.

Malzbender, D. (2005) 'Domestic electricity provision in the democratic South Africa', paper from AWIRU GiPS, University of Pretoria,SA, available from: http://www.acwr.co.za/pdf_files/01.pdf [accessed 1 January 2010].

Martinot, E. (2000) 'Making a difference in emerging PV markets: experiences and lessons from a workshop in Marrakesh, Morocco, September 2000', Global Environment Facility, Washington D.C.

Mutschler, A.S. (2007) 'Solar electricity to reach cost parity with coal-based power by 2010' [online], EDN, available from: http://www.edn.com/article/CA6432171.html [accessed 1 January 2010].

Nijaguna, B.T. (2002) *Biogas Technology*, New Age International Publishers, New Delhi.

NREL (National Renewable Energy Laboratory) (2004) 'Renewable energy in China: Township electrification program', available from: http://www.nrel.gov/docs/fy04osti/35788.pdf [accessed 1 January 2010].

Painuly, J. P. and Fenhann, J.V. (2002) 'Implementation of renewable energy technologies – opportunities and barriers', *Summary of Country Case Studies*, UNEP Collaborating Centre on Energy and Environment Riso National Laboratory, Denmark, available from: http://www.uneprisoe.org/RETs/SummaryCountryStudies.pdf [accessed 1 January 2010].

Pasternak, A. (2000) *Global Energy Futures and Human Development: A Framework for Analysis*, Lawrence Livermore National Laboratory, USA, available from: https://e-reports-ext.llnl.gov/pdf/239193.pdf [accessed 1 January 2010].

PPIAF (Public Private Infrastructure Advisory Facility) and World Bank (2007) 'PPI in developing countries in 2006, results from the PPI project database', presentation by PPIAD and The World Bank, available from: http://ppi. worldbank.org/features/Nov2007/2006Dataset.ppt#318,26,PPI in developing countries in 2006 [accessed 1 January 2010].

Powell, S. and Starks, M. (2000) 'Key drivers of improved access – service through networks', in *Energy Development Report 2000: Energy Services for the World's Poor*, pp. 44–51, ESMAP/World Bank, Washington D.C., available from: http://www.worldbank.org/html/fpd/esmap/energy_report2000/front.pdf [accessed 1 January 2010].

PowerSouth Energy Cooperative (2009) 'History of rural electric cooperatives' [online], available from: http://www.powersouth.com/news.aspx?id=119 [accessed 20 January 2010].

Practical Action (2004) 'The Upesi stove for households in Kenya – technology is only half of the story', *Energy Booklet,* Rugby, UK, available from: http://www.practicalaction.org/docs/energy/EnergyBooklet3.pdf [accessed 1 January 2010].

Program on Energy and Sustainable Development (2006) 'Rural electrification in China 1950-2004: Historical processes and key driving forces', *Working Paper 60*, Standford University, Standford, CA, available from: http://iis-db.stanford.edu/pubs/21292/WP_60,_Rural_Elec_China.pdf [accessed 1 January 2010].

Ramani, K. (2007) 'Policy research on energy access for the poor in Sri Lanka', *Policy Research Document*, Practical Action Consulting.

Reiche, K., Covarrubias, A. and Martinot, E. (2000) 'Expanding electricity access to remote areas: off-grid rural electrification in developing countries', *World Power* 2000, Isherwood Productions Ltd, London, available from: http://www.martinot.info/Reiche_et_al_WP2000.pdf [accessed 1 January 2010].

REN21 (Renewable Energy Policy Network for the 21st Century) (2007) *Renewables 2007: Global Status Report*, REN21, available from: http://www.ren21. net/pdf/RE2007_Global_Status_Report.pdf [accessed 1 January 2010].

Ruoss, D. and Taiana, S. (2000) 'A green pricing model in Switzerland: the 'Solarstrom Stock Exchange' from the electricity utility of the City of Zurich', paper from the 2nd World Solar Electric Buildings Conference, Sydney 8–10th March 2000, available from: http://www.task7.org/Public/IEA_Sydney_conference_papers/Paper_U_Daniel_Ruoss.pdf [accessed 1 January 2010].

Sager, A.D., Oliver, H.H. and Chikkatur, P.A. (2006) 'Climate change, energy and developing countries', *The Vermont Journal of Environmental Law*, Vermont Law School, available from: http://www.vjel.org/journal/pdf/VJEL10049.pdf [accessed 20 January 2010].

Saghir, J. (2005) 'Energy and poverty, myths links and poverty issues, *Energy Working Notes 4*, Energy and Mining Sector Board, World Bank, Washington D.C., available from: http://siteresources.worldbank.org/INTENERGY/Resources/EnergyWorkingNotes_4.pdf [accessed 1 January 2010].

Sanchez, T. (2006a) *Electricity Services in Remote Rural Communities The Small Enterprise Model*, Intermediate Technology Publications Ltd., Rugby, UK.

Sanchez, T. (2006b) 'Key factors for the successful implementation of stand-alone village electrification schemes in Peru', PhD Thesis, Nottingham Trent University Nottingham, UK.

SEPCO (Sustainable Energy Policy Concepts) (no date) 'Rural electrification in South Africa', available from: http://www.ises.org/sepconew/Pages/RuralEISA/document.pdf [accessed 1 January 2009].

Simões, A. F. (2006) 'Ensuring sustainable access for the poor through internal revenue generation – electricity', multi-stakeholder consultations on 'Financing Access to Basic Utilities for all', 11-12 December 2006, Brasilia, available from: http://www.un.org/esa/ffd/civilsociety/msc/utilities_brazil/AndreFelipeSimoes.ppt#256,1 [accessed 1 January 2010].

Smith, N.P.A. (1994) 'Key factors the success of village hydro-electric programmes', *Renewable Energy* 5(2), Department of Electrical Engineering, Nottingham Trent University, Nottingham, UK.

Solis, K. (2007) 'Access to GEF and CDM funds on renewable technologies for the poor', Practical Action, Bourton Hall, Bourton on Dunsmore.

Sustainable Energy Group (2008) 'Sustainable energy' [online], definition available from: http://www.sustainableenergygroup.com http://www.sustainableenergygroup.com [accessed 1 January 2010].

Turkenberg, W.C. (2000) 'Key trends is solar PV and wind energy development', presentation, Copernicus Institute for Sustainable Development and Innovation, Utrecht University, available from: libdigi.unicamp.br/document/?down=1037 [accessed 1 January 2010].

UN (United Nations) (2009) *World Economics Situation and Prospects 2009*, UN, New York, available from: http://www.un.org/esa/policy/wess/wesp-2009files/wesp2009.pdf [accessed 20 January 2010].

UNDP (United Nations Development Programme) (2002) 'Local governance for poverty reduction in Africa', *Issue paper No 2*, from the Fifth Africa Governance Forum (AGF-V) Maputo, Mozambique, 23-25 May 2002, UN and AGF-V Capital development Fund.

UNDP (2004) *The World Energy Assessment Overview*, Bureau for Development Policy, New York.

UNDP (2005) *The Human Development Report 2005, International Cooperation at a Cross Roads: Aid, trade and security in an equal world*, UNDP, New York.

UNDP and WEC (World Energy Council) (2004) *World Energy Assessment Overview 2004 Update*, Bureau for Development Policy, New York, available from: http://www.energyandenvironment.undp.org/undp/indexAction.cfm?module=Library&action=GetFile&DocumentAttachmentID=1010 [accessed 1 January 2010].

UN-Energy (2005) *The Energy Challenge for Achieving the Millennium Development Goals*, report by UN-energy, available from: http://esa.un.org/un-energy/pdf/UN-ENRG%20paper.pdf [accessed 1 January 2010].

UNFCCC (United Nations Framework Convention on Climate Change) (no date) 'Kyoto protocal' [online], available from: http://unfccc.int/kyoto_protocol/items/2830.php [accessed 20 January 2010].

Vorvate, T. and Barnes, D. (2000) 'Rural electrification in Thailand: the lessons from a successful programme', The World Bank, Washington D.C.

Wheldon, A. (2005) 'Making a business from solar home systems', case study from SELCo-India, Ashden Awards for Sustainable Energy, available from:

http://www.ashdenawards.org/files/reports/SELCO%202005%20Technical %20report.pdf [accessed 1 January 2010].

Williams, J.H. and Ghanadan, R. (2006) 'Electricity reform in developing and transition countries: a reappraisal', *Energy 31: 815-844*, Energy and Resources Group, University of California at Berkeley, Elsevier, available from: http://rael.berkeley.edu/old-site/Williams_Ghanadan_reforms_2006.pdf [accessed 1 January 2010].

WEC (2006) 'Alleviating energy poverty in Latin America: Three cities – three approaches synopsis', World Energy Council, London.

World Bank (2001) 'Peru rural electrification', *Activity Completion Report*, The World Bank, Washington D.C.

World Bank (2003a) 'Household fuel and energy use in developing countries – multicountry study', draft for discussion, 14 May 2003, prepared by Rasmus Heltberg Oild and Gas Policy Division, The World Bank, Washington, D.C.

World Bank (2003b) 'Small scale CDM projects: an overview', World Bank Carbon Finance Unit, Washington D.C., available from: http://wbcarbonfinance.org/docs/SmallScaleProcedures.DOC [accessed 20 January 2010].

World Bank (2005) 'World development indicators' [online], available from: http://devdata.worldbank.org/wdi2005/Section1_1_1.htm [accessed 20 January 2010].

World Bank (2007) 'Clean energy investment framework: progress report on the World Bank Group action plan', prepared for Development Committee Meeting October 2007, available from: http://siteresources.worldbank.org/DEVCOMMINT/Documentation/21510693/DC2007-0018(E)CleanEnergy.pdf [accessed 1 January 2010].

World Bank (2008) 'The welfare impact of rural electrification: a reassessment of the costs and benefits', *IEG Impact Evaluation*, World Bank, Washington D.C., available from: http://siteresources.worldbank.org/EXTRURELECT/Resources/full_doc.pdf [accessed 20 January 2010].

Xinhua (2006) 'Urbanization is reducing China's rural population' [online], available from people's daily website: http://english.peopledaily.com.cn/200602/23/eng20060223_245283.html [accessed 1 January 2010].

Index